03-WSM-6CO

Nutrient Farming and Traditional Removal: An Economic Comparison

by:

Donald L. Hey
Jill A. Kostel
The Wetlands Initiative

Arthur P. Hurter
Northwestern University

Robert H. Kadlec
Wetland Management Services, Inc.

Metropolitan Water Reclamation District of Greater Chicago

2005

WERF

The Water Environment Research Foundation, a not-for-profit organization, funds and manages water quality research for its subscribers through a diverse public-private partnership between municipal utilities, corporations, academia, industry, and the federal government. WERF subscribers include municipal and regional water and wastewater utilities, industrial corporations, environmental engineering firms, and others that share a commitment to cost-effective water quality solutions. WERF is dedicated to advancing science and technology addressing water quality issues as they impact water resources, the atmosphere, the lands, and quality of life.

For more information, contact:
Water Environment Research Foundation
635 Slaters Lane, Suite 300
Alexandria, VA 22314-1177
Tel: (703) 684-2470
Fax: (703) 299-0742
www.werf.org
werf@werf.org

This report was co-published by the following organizations. For non-subscriber sales information, contact:

IWA Publishing
Alliance House, 12 Caxton Street
London SW1H 0QS, United Kingdom
Tel: +44 (0) 20 7654 5500
Fax: +44 (0) 20 7654 5555
www.iwapublishing.com
publications@iwap.co.uk

Library of Congress Catalog Card Number: 2005923995

IWAP ISBN: 978-1-84339-733-5

Metropolitan Water Reclamation District of Greater Chicago, Northwestern University, The Wetlands Initiative, and Wetland Management Systems, Inc.

ACKNOWLEDGMENTS

The authors wish to acknowledge the assistance and contributions of the Metropolitan Water Reclamation District of Greater Chicago, specifically, Joseph Sobanski, Nicholas Venuso, Adam Gronski, and Mery Zuniga from the Engineering Department and Richard Lanyon and Bernard Sawyer from the Department of Research & Development. David Reardon, who has studied nutrient control technology for many years, was kind enough to share his knowledge and experience and assisted with formulating and reviewing the cost estimates for the water reclamation plants.

Report Preparation

Principal Investigator:
Donald L. Hey, Ph.D.
The Wetlands Initiative

Project Team:
Arthur P. Hurter, Ph.D.
Northwestern University

Robert H. Kadlec, Ph.D.
Wetland Management Services, Inc.

Jill A. Kostel, M.S.
The Wetlands Initiative

Project Subcommittee

Water Environment Research Foundation Staff

ABSTRACT AND BENEFITS

Abstract:

Wastewater treatment plants (WWTPs) would be well served by large-scale treatment wetlands to control nutrients. Based on the U.S. Environmental Protection Agency's recommended nutrient criteria, an economic analysis of conventional treatment vis-à-vis treatment wetlands was conducted. Conventional technology was represented by biological nutrient removal; treatment wetlands were represented by restored riverine ecosystems. Cost functions for both technologies were constructed and are presented. These functions were used to compute and compare total, average, and marginal costs. In addition, they were used to define supply and demand curves for nutrient credits. As required by these analyses, the necessary wetland area was estimated.

For explanatory purposes, a case study was structured using the seven WWTPs of the Metropolitan Water Reclamation District of Greater Chicago. Between 189,000–322,000 acres of treatment wetlands would be necessary to treat the required nutrients of these plants, depending on the effluent limit. In turn, wetland-based removal provides a savings of 51–63% of the annual cost of conventional treatment with the savings increasing to 76–78% if surplus nutrient credits generated by the treatment wetland can be sold.

Nutrient farming is a proposed concept for constructing and operating treatment wetlands. Such farms would offer the organizational and financial means for reliably producing nutrient credits. Suitable lands can be found within local floodplains, for example along the Illinois River. However, until large-scale nutrient farms are in existence and operating experience is gained, there will be reluctance to use this technology. Pilot projects, in various ecoregions, are necessary to establish optimal design and operating procedures and to demonstrate the economic efficiency of treatment wetlands.

Benefits:

- Offers cost equations for nutrient removal using conventional wastewater technology and treatment wetlands.
- Provides a nutrient removal model—and associated parameters—for treatment wetlands.
- Demonstrates the savings in capital and operating costs that treatment wetlands can provide.
- Proposes the development and use of nutrient farms as the organizing concept and financial structure for the construction and operation of treatment wetlands.
- Recommends the establishment of pilot projects (nutrient farms).

Keywords: nutrient management, treatment wetlands, nutrient farming, cost equations, present value and marginal cost

TABLE OF CONTENTS

LIST OF TABLES

LIST OF FIGURES

LIST OF ACRONYMS

BNR	biological nutrient removal
BOD	biological oxygen demand
EA	effective area
FSW	free surface water
LIS	Long Island Sound
MGD	million gallons per day
MWRDGC	Metropolitan Water Reclamation District of Greater Chicago
NH_4-N	ammonia-nitrogen
NO_3-N	nitrate-nitrogen
N_{org}	organic nitrogen
NPS	nonpoint source pollution
O&M	operating and maintenance
TMDL	total maximum daily load
TKN	total Kjeldahl nitrogen
TN	total nitrogen
TP	total phosphorus
UMRB	Upper Mississippi River Basin
U.S. EPA	United States Environmental Protection Agency
WRP	water reclamation plant
WWTP	wastewater treatment plant

EXECUTIVE SUMMARY

Point source nutrient control is on the horizon. The governors of Wisconsin and Minnesota have both agreed to reduce by 30% the total nitrogen load (nonpoint and point source combined) discharged to their state's water bodies. The U.S. Environmental Protection Agency (U.S. EPA) has set the nutrient criteria, and the individual state agencies are beginning the standard setting process. In Ecoregion VI, whether by total maximum daily load allocations or other regulatory means, the ultimate standards will be near the U.S. EPA's current ambient water quality criteria of 1.16–3.26 mg/L and 0.063–0.118 mg/L for total nitrogen (TN) and total phosphorus (TP), respectively (U.S. EPA, 2000). Although there are advanced treatment technologies that can reduce the nutrient load discharged from municipal and industrial point sources, there are no technologies, within reasonable economic limits, that will reduce the loads of phosphorus and nitrogen sufficiently to meet the criteria. Physical, chemical, and biological nutrient controls are capital intensive and require substantial annual operating and maintenance costs.

To explore the economic relationship between wastewater treatment technology and treatment wetlands and to quantify the magnitude of wetland area needed, a case study was developed. This study has examined the seven water reclamation plants (WRPs) owned and operated by the Metropolitan Water Reclamation District of Greater Chicago (MWRDGC) and the potential use of restored wetlands within the Illinois River watershed. To compare these two treatment systems, cost functions were developed for the seven WRPs and for the treatment wetlands. These functions were based on the actual operating parameters and conditions of the WRPs and wetlands. Biological nutrient control, specifically the five-stage Bardenpho (with methanol addition), was assumed to be the advanced treatment technology used in upgrading the WRPs. The MWRDGC formulated the capital cost for a single WRP then applied the capital cost equations and design to its other WRPs.

For the treatment plants, the derived monthly cost functions covered a treatment capacity of 3.4 to 1200 million gallons per day. Seasonal cost functions for nitrogen and phosphorus removal were developed based on land, pumping and restoration costs, and nutrient concentrations within the Illinois River (the effluent receiving stream). The outfall load from the WRPs determined the area of wetlands needed to remove the excess nutrient loads for two sets of WRP effluent standards (3.0 mg/L of TN and 1.0 mg/L TP; 2.18 mg/L TN and 0.5 mg/L TP).

Given the physical, chemical, and biological processes governing natural systems, the winter months reflect the least efficient wetland treatment and establish the minimum land area necessary to treat the monthly MWRDGC demand. This wetland area is much larger than that required for treatment during the spring and summer seasons. As a result, surplus wetland acreage that could produce nitrogen and phosphorous credits for other emitters, such as power plants, would be available during the warmer months. On the other hand, during periods when less nutrient removal is needed, the operating conditions and costs at treatment wetlands could be curtailed.

On the basis of the cost functions, the load-weighted average and marginal costs for each treatment technology were determined. The average costs for the seven WRPs are $49,500/ton and $8,130/ton for TP and TN, respectively, for the higher criteria of 1.0 mg/L and 3.0 mg/L for TP and TN, respectively. The nitrogen cost estimate compares favorably with seven of the larger treatment plants tributary to the Long Island Sound, where the load-weighted average cost for TN removal was calculated to be $6,870/ton. In comparison, the load- and seasonal-weighted average costs for wetland treatment systems are $2,220/ton for TP and $2,250/ton for TN. Because the cost functions were determined to be linear, Y ($) =aX (tons of nutrient removed) + b, the average cost, Y/X = (a + b /X), is a declining function. Consequently, the greater the load treated the less the average cost. Still, within the conditions of the case study, the average cost of wetlands is substantially less than that for wastewater treatment plants.

In this study, the marginal cost, or the price of removing the last ton of nutrient, is the cost value. Because the marginal cost is the first derivative of the cost function, "a" equals the marginal cost. After the first ton of removal occurs, the average cost curve declines whereas the marginal cost remains the same. Still, considering the marginal cost of the WRPs versus wetlands, wetlands have a lower cost by almost an order of magnitude. The load-weighted marginal costs for wetlands are $1,830/ton and $1,930/ton for TP and TN, respectively, whereas the load-weighted marginal costs for the seven WRPs are $16,000/ton and $3,410/ton for TP and TN, respectively.

The overall annual cost of operating treatment wetlands is 51–63% less than the cost of constructing and operating conventional wastewater treatment plants. The savings could be even greater, 76–78%, if secondary markets for the surplus nitrogen and phosphorus credits could be developed.

Despite the seasonality of nutrient removal by wetlands, wetlands proved to be the more efficient means of nutrient control. In addition, wetlands would provide other valuable benefits to the environment: flood control, wildlife habitat, recreational opportunities, and public education. Given this potential, the next step in this evaluation would be the development of pilot projects distributed across the country by ecoregions and climatic zones. These pilot projects would verify this study's cost functions and research the most efficient operating regimen for the treatment wetlands. These pilot projects, at the same time, would afford environmental engineers and wastewater treatment plant operators the opportunity to become familiar with and learn how to manage treatment wetlands.

CHAPTER 1.0

INTRODUCTION

1.1 Purpose and Objectives

The purposes of this study were to assess the economic feasibility of using large-scale, restored wetlands to assist wastewater treatment plants (WWTPs) in meeting the U.S. Environmental Protection Agency's (U.S. EPA) published criteria for nutrients, specifically, total nitrogen (TN) and total phosphorous (TP), and to evaluate the size of treatment wetlands needed. The assessment compares the cost of nutrient control by advanced wastewater treatment technology to that of wetland treatment technology. The comparison was based on several economic factors: annual operating costs, average costs, marginal costs, and present value. The economic analyses were defined and quantified based on a case study of the water reclamation plants (WRPs, which are equivalent to WWTPs) owned and operated by the Metropolitan Water Reclamation District of Greater Chicago (MWRDGC). In turn, these treatment plants were placed in the context of their watershed—the upper Illinois River—for the assessment of the cost to restore and operate wetlands as treatment systems. Comparing the costs of these two technologies was the first purpose of this study.

The second purpose was to estimate the necessary area of treatment wetlands needed to meet the nutrient removal demand of the MWRDGC. For traditional wastewater treatment, land area is minimized at the expense of concrete, steel, energy, and labor. In contrast, wetland treatment is performed at the expense of land. Although the land requirements for traditional wastewater treatment are well established, little is known about the land requirements for a given nutrient load reduction under the rigorous, required operational conditions: 365 days a year, seasonal influences and flood inundation. These conditions have not been fully tested despite numerous studies (e.g., Hey et al., 1994; Kadlec, in press; Mitsch et al., 1999). Except for flooding, this study considers the entire range of critical conditions.

1.2 Nutrient Water Quality Criteria and Standards

For environmental reasons, the U.S. EPA issued criteria for the concentration of TN and TP in streams and rivers of the United States (U.S. EPA, 2000). The criteria, expressed in terms of concentrations, vary from ecoregion to ecoregion across the country. For the Midwestern states (Minnesota, Iowa, Missouri, Illinois, and Wisconsin) located in Ecoregion VI (Corn Belt and Northern Great Plains), the recommended criteria are 2.18 mg/L for TN and 0.076 mg/L for TP. These criteria levels are considerably less than the concentrations observed in the Illinois River today. The concentration of TN found in the Illinois River near Peoria often exceeds 6.8 mg/L; TP concentrations often exceed 1.5 mg/L. State agencies, authorized tribes, and territories are mandated by the U.S. EPA to write and enact water quality standards or to adopt the recommended water quality criteria for the relevant ecoregion.

The U.S. EPA's concern over increasing nutrient concentrations relates primarily to the ecological health of the nation's waters, although occasional concentrations of nitrate-nitrogen (NO_3-N) may exceed the drinking water standard of 10 mg/L, which could lead to "blue baby syndrome." Less extreme concentrations of nitrogen, yet considerably above the proposed criteria, are thought to cause cancer in humans (Weyer et al., 2001). In addition to nitrate, elevated phosphorus concentrations—typically a growth-limiting factor for freshwater algae—can be responsible for nuisance algal blooms. Indicators of nutrient overenrichment include an overabundance of algae and macrophytes, low dissolved oxygen concentrations, fish kills, and a depletion of desirable flora and fauna.

The effect of high nitrogen loads reaching the coastal waters is a global concern. Excessive nitrogen loads, conveyed by streams and rivers, are thought to cause hypoxia, or low dissolved oxygen, in the coastal ecosystems. In the Gulf of Mexico, for example, the overenrichment of nitrogen, mainly NO_3-N, increases algal production. The decomposition of the dead algae leads to low dissolved oxygen concentrations (<2 mg O_2/L) in the water column. During the late spring and early summer months, the uncharacteristically low dissolved oxygen levels force more mobile organisms (e.g., fish) to flee and kill less mobile organisms (e.g., shrimp). In either case, commercial fisheries are affected, as are biodiversity and the ecological health of the aquatic ecosystem. Before 1993, the hypoxia zone in the Gulf of Mexico averaged 3000–3500 sq mi; however, the extent of the hypoxic area ranged from 6100 sq mi to 7700 sq mi between 1993–1999 (Rabalais et al., 1999; Rabalais, 2004).

Although point source discharges, soil depletion, and atmospheric deposition contribute to the nutrient loading in the Mississippi River Basin, the majority of the aqueous nitrogen and phosphorus found in streams and rivers comes primarily from agricultural activities (Goolsby et al., 1999). The use of commercial fertilizers, the application of manure, and the production of legumes (e.g., soybeans) all contribute to increased concentrations of nitrogen. The efficient drainage systems now employed across agricultural landscapes hasten the movement of nitrogen and phosphorous into streams and rivers and reduce the ability of a tributary drainage system to retain or recycle these constituents (Hey, 2002). For example, in the less-urban areas of Illinois, where agriculture is the dominant land use, surface runoff and subsurface tile drainage is the most significant source of nitrogen (86%) and phosphorus (67%) loading, whereas, one-fifth of the nitrogen loading and two-thirds of the phosphorus loading in the Illinois River is from urban municipal and industrial effluent discharges (David & Gentry, 2000).

To restore the health of our nation's receiving waterbodies, the load reduction is enormous. To reduce the nitrogen load reaching the Gulf of Mexico to the pre-1900s concentration of 0.8 mg/1, the NO_3-N load in the Illinois River would need to be decreased by 80%, or 101,000 tons per year (Hey, 2002). If these load reductions are to be achieved, both point and NPS contributions must be greatly reduced. Although it might be argued that the criteria are exceedingly stringent, the need for a reduction in nutrient loads is obvious both in terms of the ecological integrity of the inland, freshwater ecosystems as well as that of the Gulf of Mexico.

1.3 Nutrient Removal Technologies

From a regulatory perspective, the easiest place to start the reduction of nutrients is with point source dischargers (i.e., municipal and industrial). The dischargers are regulated under the Clean Water Act, whereas nonpoint sources (NPS), such as agricultural runoff, are not regulated. Clearly, agricultural runoff is the largest source of nitrogen and phosphorous, but it defies conventional treatment because of its wide geographical distribution and highly variable flow and nutrient concentrations. On the other hand, point sources are easily identified and their treatment technologies are well defined and tested, but expanding and upgrading the existing treatment plants will be extremely expensive. Perhaps the proposed technology, nutrient farming, could meet the nutrient management needs of both point sources and NPS.

If the nutrient farming strategy proves technically, environmentally, and economically effective for point source control, the point source application could establish the principles and examples to lead the way to effective NPS control. But before the nutrient farming strategy is applied to the point source case, its effectiveness must be proven. This will require, ultimately, several large-scale pilot projects. First, the theoretical basis for these projects will need to be established. To start this theoretical development, the analyses set out in this report establish the economic principles for nutrient farming by comparing the economics of nutrient farming with conventional treatment.

1.3.1 Conventional Treatment

Any reduction in biological oxygen demand (BOD), total suspended solids, nitrogen, and phosphorus beyond the levels typically reached by secondary treatment is referred to as advanced treatment. Several advanced treatment technologies have been developed and used to lower nitrogen and phosphorus discharges: biological nutrient removal (BNR), chemical removal (metal salts or lime), and physical processes (membrane separation, ionic exchange, reverse osmosis, air stripping, filtration, etc.).

Physical and chemical process (e.g., air stripping, ionic exchange) can remove nitrogen; however most systems used biological treatment for the removal of nitrogen. The biological treatment of nitrogen involves creating conditions for nitrification and denitrification either in suspended growth or fixed-film systems. Nitrification is an aerobic process in which bacteria oxidize ammonia to nitrate; denitrification is the conversion of nitrate to nitrogen gas. Although nitrification occurs in secondary treatment systems (e.g., activated sludge or trickling filters), existing systems can incorporate denitrification by including an anaerobic process after effluent nitrification. Because denitrifying bacteria depend on organic carbon for heterotrophic growth, methanol is often added.

Although the final product in nitrogen removal can be a gas (N_2), phosphorus removal depends on the conversion of the phosphate species to a particulate form and its subsequent removal by a solid removal process (i.e., sedimentation or filtration). Phosphorus can be removed by chemical precipitation using metal salts (e.g., ferric chloride, alum) or by cellular uptake and incorporation into bacteria biomass in excess of typical cell requirements. Although chemical addition is operationally simple and can achieve levels less than 0.1 mg/L TP (Reddy, 1997), the use of chemicals is expensive and results in an increase in sludge production.

It is generally accepted that biological nitrogen removal is more economically feasible than physical/chemical treatment methods; however, this is not necessarily the case for biological phosphorus removal, because the cost of enhanced biological phosphorus removal depends on influent water characteristics (BOD and TP) and removal method as specific environmental conditions must be created to enable specific bacteria to uptake large quantities of phosphorus (Reardon, 1994).

Advanced treatment technologies can provide a reliable effluent under design flow conditions; however the use of this technology has ancillary issues. The necessary plant expansions and improvements will increase biosolid production and the issues surrounding biosolid disposal. The energy costs of a wastewater treatment facility are a major cost factor, typically 20–40% of the total annual operating budget (Water Environment Research Foundation, 2004). The electrical energy usage will increase, depending on the advanced treatment technology employed, as pumping demands will increase for aeration, nitrate recycling, and sludge return cycles. In turn, greater amounts of fossil fuel will need to be combusted to produce the energy. These carbon emissions, as well as the volatilization of methane and nitrous oxide from organic biosolid degradation, will increase the greenhouse gases emanating from the wastewater treatment process.

1.3.1.1 Biological Nutrient Removal

For this study, the BNR five-stage, modified Bardenpho (with methanol addition) treatment process was selected for the case study because the MWRDGC is investigating the use of this technology. The five-stage Bardenpho uses anaerobic/anoxic/aerobic-activated sludge environments, with the methanol supplying the carbonaceous energy for denitrification. This process can produce, after clarification and filtration, an effluent of 3.0 mg/L TN and 1.0 mg/L TP. These effluents are well above the U.S EPA's Ecoregion VI criteria. If the Illinois EPA does not enact nutrient standards that are less stringent than the U.S. EPA criteria, additional treatment methods of processes will be necessary (e.g., chemical addition for enhanced phosphorus removal).

1.3.1.2 MWRDGC Case Study

The MWRDGC currently operates seven WRPs. The plants vary in design flow capacities, in treatment operation, and in the amount and type of nitrogen [NO_3, NO_2, and total Kjeldahl nitrogen (TKN)] and phosphorus loadings (Table 1-1). The design flows of the MWRDGC plants were from 3.4 million gallons per day (MGD) to 1200 MGD—the largest WRP in the world.

Table 1-1. TN and TP Influent and Effluent Flow-Weighted Mean Concentrations in 2002.

WRP	Design Flow (MGD)	Average Daily Flow (MGD)	TN		TP	
			Influent (mg/L)	Effluent (mg/L)	Influent (mg/L)	Effluent (mg/L)
Lemont	3.4	2.17	42.6	16.0	5.5	2.69
Hanover Park	12	8.18	45.5	11.0	5.3	2.79
John E. Egan	30	23.8	46.7	15.4	7.0	3.14
James C. Kirie	72	33.4	35.5	7.5	4.1	0.85
North Side	333	250	30.5	9.8	3.4	1.33
Calumet	354	237	35.6	9.2	7.1	3.20
Stickney	1,200	683	47.0	10.8	6.2	1.25

All seven plants operate conventional activated sludge treatment processes with nitrification to reduce ammonia and its toxic effects on receiving streams. Three of the plants—Egan, Hanover Park, and Kirie—have a tertiary filtration stage before chlorination, whereas, secondary treatment (settling or clarification) is the final treatment stage for the four other plants. The percentage of nutrients removed, based on mass, varies among the plants. Among the plants, the average TN removal efficiency varies only modestly: The average for all seven plants is 72% (±6 %). The removal of TP varies more widely among the plants, from 48% (Hanover Park) to 80% (Stickney).

Currently, there are no full-scale advance treatment processes for nutrient removal operating at any of the plants. Relative to the U.S. EPA nutrient criteria recommendations for Ecoregion VI, the WRPs are currently discharging effluents that are 3.4 to 7.3 times higher for TN and 11 to 43 times higher for TP. Consequently, the MWRDGC will need to upgrade its WRPs to achieve most any reasonable nutrient standard.

1.3.2 Wetland-based Treatment

1.3.2.1 Nutrient Removal in Wetlands

If treatment wetlands are properly designed, constructed, and managed, their naturally occurring functions should provide an economically sustainable means for reducing the nutrient loads found in streams and rivers. In terms of improving water quality and erosion control, the value of wetlands is found in the inherent biogeochemical and physical processes that remove, transform, and sequester nutrients such as nitrogen, phosphorus, and carbon. These processes must be taken into consideration when designing treatment wetlands for nutrient removal.

Influent nutrients, whether dissolved or associated with particulate matter, are removed from the water by a combination of physical, chemical, and biological processes driven by the hydrologic forces and catalyzed by the extant soils and plants. These processes may be affected by temperature, pH, presence or absence of oxygen, microbial communities, substrate carbon characteristics and concentrations, nutrient availability in soils and sediments, and substrate properties such as texture and structure (Reddy & Patrick, 1984; Crumpton et al., 1994; Phipps, 1997). Physical transport processes include particulate settling and resuspension, diffusion of dissolved species, litterfall, plant uptake, sorption of soluble nitrogen on substrates, and volatilization (Kadlec & Knight, 1996).

The transformation of nutrient species in wetlands occurs by the processes of assimilation (plant and bacterial uptake of phosphorus and nitrogen species), mineralization (ammonification), nitrification, denitrification, and nitrogen (N_2) fixation. Three forms of nitrogen are important as respeciation takes place: nitrate and nitrite (oxidized), ammonia (reduced), and organic nitrogen (N_{org}), which is determined as the difference between measurable total TKN and ammonia-nitrogen (NH_4-N). As emergent macrophytes produce large amounts of detrital carbon (e.g., Murkin, 1989), the high carbon-to-nitrogen ratio reflects a demand. The demand for additional nitrogen is met through assimilation or fixation (Bowden, 1987). Nitrogen assimilation is the biological process that converts inorganic nitrogen species, mainly NH_4-N and NO_3-N, into organic compounds used for cell and tissue growth (Kadlec & Knight, 1996). During growing seasons (spring and summer), plants uptake nitrogen; during the winter months, nitrogen can be released after senescence. In addition to plants and algae, microorganisms can assimilate nitrogen for growth. The energetically preferred nitrogen species, NH_4^+, is readily incorporated by many autotrophs and microbial heterotrophs (Kadlec & Knight, 1996). Nitrogen fixation is an adaptive process that provides nitrogen for organism growth in conditions that are devoid of nitrogen. Atmospheric nitrogen gas that is diffused into solution is reduced to N_{org} by autotrophic and heterotrophic bacteria, cyanobacteria, and higher plants (Kadlec & Knight, 1996). Typically, this process is not observed in treatment wetlands that receive nitrogen-laden waters.

The overall effectiveness of wetlands as nitrogen sinks depends on NO_3-N and the capacity of wetlands to remove nitrate by denitrification. Denitrification is the biological reduction of NO_3-N to gaseous nitrogen species (i.e., N_2, nitrous oxide, nitric oxide) under anaerobic conditions. Wetlands that receive significant and sustained nitrate loadings have demonstrated high rates of denitrification with more than 80% of the externally loaded nitrate lost through denitrification (Moraghan, 1993; Crumpton et al., 1994). Water depth in free surface water (FSW) wetlands needs to be controlled, as denitrification is limited in systems with low NO_3-N availability in the sediments because of the low redox conditions associated with high water levels.

Although there is not a dissipation pathway for phosphorus removal as there is for nitrogen (via denitrification), significant quantities of phosphorus are deposited, adsorbed, or used in wetlands (Richardson, 1985; Johnston, 1991; Walbridge & Struthers, 1993). Mainly the physical process of sedimentation removes particulate phosphate. Emergent and submerged vegetation facilitate the sedimentation of particles as the stems and leaves enhance the deposition of suspended sediments and phytoplankton by providing friction in flowing waters (Johnston, 1991). The other dominant physical process for phosphorus removal is the sorption of soluble phosphorus. The sediment associated phosphorus can become available to the plant root system via desorption, reversed chemical binding, and porewater diffusion. Orthophosphate, the predominant inorganic form of phosphorus (PO_4-P), accumulates readily in sediments and vegetation by adsorption and precipitation reactions (chemical bonding) or as a result of direct assimilation (biological uptake), respectively. For a range of wetland soils/sediments, phosphorus adsorption (aluminum and iron oxides and hydroxides) and phosphorus precipitation (aluminum, iron, and calcium phosphates) are thought to be the two processes by which the majority of phosphorus is removed for long-term storage (Richardson et al., 1988; Cooke, 1992; Walbridge & Struthers, 1993).

Subsurface soils and associated plant tissues sequestered the majority of the phosphorus in natural and constructed wetlands. The short-term phosphorus storage and nutrient cycling provided by wetland biota are important processes that regulate net phosphorus retention in wetlands (Richardson & Craft, 1993; Kadlec & Knight, 1996); however, assimilation of phosphorus by biota (macrophytes, algae, bacteria) can be minor in comparison to the fraction retained by the sediments (Kim & Geary, 2001). The above-ground macrophyte biomass is generally replaced between 1 and 2 times per year in northern environments (Mitsch & Gosselink, 1993); however, overall plant biomass (live, standing dead, and litter) remains constant within a system across the seasons (Kadlec & Hammer, 1988).

Although plant assimilation can incorporate a significant fraction of the TN and TP load in a wetland, it is a seasonal phenomenon. There are seasonal releases of assimilated nutrients back into the wetland system during the fall and winter months as nutrients are leached out from macrophyte tissues during senescence. The decomposition of the macrophytic leaf litter and microbiota is the primary means of accruing new sediment and soil material. The net accumulation of organic material and sediment particulates is 1–10 mm/yr (Kadlec & Knight, 1996; Reddy et al., 1991). The accretion of new material is necessary for the continual removal of phosphorus and serves as the only sustainable storage mechanism for phosphorus removal.

Several treatment wetlands have been successfully removing nutrients from rivers receiving urban runoff and municipal effluent. The Prado Basin Wetland, a 465-acre wetland located in Riverside County, California, has been receiving waters (up to 100 ft^3/s) from the Santa Ana River since 1992. Primarily through denitrification, this wetland system has removed 20 tons NO_3-N per month and has recorded reductions in NO_3-N concentrations from 10 mg/L to less than 1 mg/L during the summer months (Lund et al., 2000; Reilly et al., 2000). The San

Joaquin Marsh has been developed and used to treat urban runoff from the San Diego Creek Watershed. The 500-acre wetland, which receives 3.25–6.5 MGD from San Diego Creek, removes about 15 tons of nitrogen per year (Irvine Ranch Water District, http://www.irwd.com).

Sound design and management practices will be necessary for optimizing nutrient removal throughout the year. Research will be needed to establish best hydraulic conditions (e.g., flow rates, depth, wetted surface, and residency time), soil characteristics, and plant community structures (Nungesser & Chimney, 2001). The effluent of treatment wetlands can meet regulatory standards; however, it depends on wetland size and nutrient loading rates. The removal processes have a finite capacity; therefore, once the capacity of these processes is exceeded, removal efficiency declines. Based on a survey of 141 treatments systems, the majority of the wetlands achieved effluent concentrations of 0.1–0.2 mg/L TP (Kadlec, unpublished data). Research and monitoring will be necessary to determine the nutrient load that can be safely assimilated and used, given the design conditions, without a negative effect on critical ecological structures and functions (Keenan & Lowe, 2001). A robust, diverse wetland should be able to respond positively to new and variable influxes of nutrients by readjusting storage capacities, pathways, and structures (Kadlec & Knight, 1996).

1.3.2.2 Nutrient Farms

Nutrient farming is a strategy for developing and operating treatment wetlands. The term farming comes from the activity of growing wetlands to harvest nutrients and, as a related benefit, suspended solids and sediment. These are ecological functions that presettlement wetlands once performed. Wetlands were abundant in regions such as the Upper Mississippi River Basin (UMRB). In the past 200 years, three states (Illinois, Iowa, and Missouri) in the UMRB have drained 85–90% of their wetlands (Dahl, 1990). More than four million acres of wetlands have been destroyed in the Illinois River watershed (Dahl, 1990). Today, there is a greater appreciation of the important functions that wetlands played in this landscape, especially in terms of ability of wetlands to improve water quality.

Nutrient farming is a strategy intended to provide the economic incentive for restoring a small, but critical, portion of the lost wetlands and, at the same time, reducing the nutrient load coursing this nation's streams and rivers. The strategy involves converting, for example, low-lying corn and soybean fields to wetlands. Nutrient-laden waters then would be diverted to and detained on the restored landscapes. The diverted waters would be returned to the stream after a portion of the nutrient load has been removed. Through this diversion process and with nutrient farms distributed along streams and rivers, the requisite nutrient load would be removed and the standard achieved.

Restoring riverine wetlands is suggested as opposed to end-of-pipe constructed or artificial wetlands—manmade systems often situated in areas where wetland ecosystems never occurred. On the other hand, the use of large-scale treatment wetlands, or nutrient farms, is not feasible in highly developed, urban areas where land availability is limited and land costs are high or in highly productive agricultural areas. Locating the nutrient farms in downstream reaches allows the development and operation of larger farms, which can more effectively target larger flows (Crumpton, 2001). These restored wetlands could accommodate all nutrient-laden watershed waters originating from point source and NPS. In cases where the nutrient farm is located and operated some distance from the point of nutrient discharge, as is the case in this study, regulatory agencies will need to identify stream reaches where nutrient transport would be the designated use. These reaches will likely be in highly modified channels (e.g., the Illinois Waterway) and removed from primary contact and drinking water uses.

Developing and operating a nutrient farm would be fairly simple. First, drainage controls would need to be established. This might involve blocking drain tiles, constructing inflow and outflow weirs and, even, building pumping stations, which is contemplated in this study. The farm operator would record, on a daily basis, the influent and effluent loads and compute the difference. The mathematical result would represent the earned nutrient credits. The credits would be certified by the state, based on the operator's Daily Monitoring Report and occasional site verification. Once the operator has received state certification, the credits could be sold on the open market or applied to an existing sales contract.

The operator may be an individual, association (e.g., drainage district), corporation, or government agency (e.g., water reclamation district). The buyers could be other farmers, municipal and industrial dischargers, and those emitting nitrogen to the atmosphere, which ultimately falls back to the land's surface, such as power plant and automobile owners. The relationship between seller and buyer would be left to their discretion except for the allowable nutrient load being emitted and the load removed. These conditions remain the purview of the regulating state agency.

Certainly, nutrient farming will require the development of markets and rules of governance. Although neither is well defined at this point, they are clearly necessary to protect human health and the environment. They would provide for the exchange and certification of nutrient credits, generate farm income and capital for expanding and improving nutrient farms (i.e., increasing wetland resources), optimize the allocation of natural resources and, ultimately, internalize the cost of agricultural production.

Once state, tribal, and territory agencies enact water quality standards, the total maximum daily load (TMDL) for nitrogen and phosphorus will establish the cap on nutrient loads. Point, and, one day, nonpoint, sources will be required to reduce, or at least not increase, their emissions of nutrients without buying nutrient credits. Nutrient farm industry would be responsible for removing the requisite load to maintain the relevant TMDL and meet the state standard.

1.3.2.3 Ancillary Benefits and Environmental Issues with Nutrient Farms

Although the primary goal of nutrient farming is to improve water quality both in local watersheds and in distant coastal ecosystems, the restored wetlands will provide important ancillary benefits unavailable through the use of traditional treatment technology. These benefits include reduced flood damages; removal of suspended solids, heavy metals, pesticides, water-borne pathogens; increased wildlife habitat and biodiversity; hiking, fishing, hunting, and bird-watching recreational opportunities; alternative source of farm income; and climate moderation caused by carbon sequestration. The restoration of wetlands will offset the loss of this type of ecosystem, which is occurring at a rate of approximately 58,500 acres per year in the contiguous United States because of agricultural and urban, residential development (Dahl, 2000).

Positioning treatment wetlands adjacent to rivers and systems will allow for the conveyance and storage of floodwaters through and on wetlands during periods of high flow. The storage of floodwaters may temporarily impair treatment potential; however, operational protocols can be developed and implemented to accommodate the floodwaters and the decrease in nutrient removal efficiency. Design and operation modifications have been accomplished in existing floodplain treatment wetlands (Brunner et al., 1992). Further, controlled input and release of floodwaters can minimize damage to wetland control structures, pumps, and vegetation.

At most locations, the restored wetlands will add needed habitat for waterfowl and wildlife otherwise not present in the immediate vicinity or local watershed. Waterfowl damage on very large wetlands is not deemed to be serious and control measures are thought to be undesirable. There are many major waterfowl (refuge) wetlands in North America, and these do not typically require management of herbivory. Many large "natural" wetlands function in an oscillatory balanced mode without ecosystem destruction. Bottom-foraging fish, such as carp, will likely be present sooner or later in the wetlands. Their presence can create disturbed sediments and turbid waters, which may impair nutrient removal processes. Depending on the effects on nutrient removal, control measures may need to be considered. It is the presumption in this analysis that hydraulic measures could serve this purpose (e.g., water draw-down and freezing).

As with other shallowly flooded habitats, wetlands can potentially provide a breeding habitat for mosquitoes. Mosquitoes present a biting nuisance and a vector of diseases (e.g., West Nile virus) to livestock and humans; however, the actual occurrence of mosquito-related health problems associated with treatment wetlands is rare, especially in northern climates (Kadlec & Knight, 1996). In a properly designed and managed wetland, mosquito populations can be controlled by natural aquatic predators (e.g., dragonfly and damselfly larvae), maintaining constant water levels and avoiding high organic loadings. In addition, there is no known instance of an odor problem related to wetlands treating such low-strength wastewater.

CHAPTER 2.0

CONVENTIONAL TREATMENT COST ANALYSIS

2.1 BNR Nutrient Load Removal Calculations

Because the Illinois EPA water quality standards for nitrogen and phosphorus have not been enacted, the total cost equations were structured to represent a range in TN or TP effluent limit. The modified five-stage Bardenpho achievable effluent limits of 3.0 mg/L TN and 1.0 mg/L TP were considered the upper permissible water quality limits for the calculation of the mass of nutrients needed to be removed from the treatment plant effluent. In addition, the tons of TN and TP required to be removed to achieve stricter effluents limits (2.18 mg/L TN and 0.5 mg/L TP) were calculated. These limits were chosen based on the Aggregate Ecoregion VI criterion for TN (2.18 mg/L) and the estimated achievable TP limit for the MWRDGC's WRPs (0.5 mg/L). For the case study, the daily influent and effluent concentrations for each plant in 2002 were used to determine the relevant loads.

The load of TN and TP that needs to be removed by advanced treatment processes in order for each WRP to meet the design effluent limits was calculated by subtracting the mass load associated with the targeted effluent limit from the existing effluent load. The difference was termed the excess load. Mass load, in lb/day, was determined by multiplying the existing effluent concentrations or effluent limit value (mg/L) by the daily flow and a unit conversion factor of 8.34 [(lb l)/(mg 10^6 gal)]. The excess load was calculated on a daily basis for each WRP.[1]

Using 2002 data, the annual nutrient discharge, permissible loads and excess loads for the seven plants are presented in Tables 2-1 (TN) and 2-2 (TP). For all seven plants, the TN loads that need to be removed to achieve the 3.0 mg/L and 2.18 mg/L effluent limits are 13,700 and 15,300 tons, respectively. Approximately 1270 and 2210 tons of TP need to be removed to achieve the proposed 1.0 mg/L and 0.5 mg/L TP effluent limits, respectively. Only the Kirie plant achieved the 1.0 mg/L TP limit. However, this plant was not operating at full capacity. In addition, the Stickney and North Side WRPs met the upper effluent limit for TP on several days in 2002. The excess monthly loads needed to be removed to achieve the 3.0 mg/L TN and 1.0 mg mg/L TP effluent limits were used in developing the total cost equations for the seven individual WRPs.

---

[1] Since the effluent data sets for each WRP were not complete for all the nutrient species of interest because of sampling protocols (5 days a week) or missing test samples, several daily concentrations had to be estimated. Single or consecutive missing data points were estimated based on the average percent removed from the raw influent water since the percent removed between influent and effluent is fairly consistent within a specific WRP. The average of the percent removed was calculated from the 5 days before and 5 days after the missing data point if the nutrient species was sampled 7 days a week. Otherwise, the average was calculated from the three days before the missing data point and the next three recorded measurements. The Stickney WRP did not have 20 consecutive TKN (TN = TKN + NO_3-N + NO_2-N) measurements for any treatment stage including the raw influent; therefore, the outfall concentrations were estimated from a correlation between effluent TKN concentration and the recorded daily flow.

Table 2-1. Total Annual Nitrogen Load for 2002.

WRP	Influent Load (tons)	Existing Outfall Load (tons)	3.0 mg/L Effluent Limit		2.18 mg/L Effluent Limit	
			Permissible Limit (tons)	Excess (tons)	Permissible Limit (tons)	Excess (tons)
Lemont	141	53.0	10.0	43.0	7.10	45.9
Hanover Park	566	137	37.4	99.4	26.6	110
John E. Egan	1690	558	109	449	77.4	481
James C. Kirie	1800	380	152	227	109	271
North Side	11,600	3,730	1,140	2,580	813	2,910
Calumet	12,900	3,310	1,080	2,230	773	2,540
Stickney	49,400	11,200	3,160	8,100	2,250	9,000
Total	78,100	19,400	5,690	13,700	4,060	15,300

Table 2-2. Total Annual Phosphorus Load for 2002.

WRP	Influent Load (tons)	Existing Outfall Load (tons)	1.0 mg/L Effluent Limit		0.50 mg/L Effluent Limit	
			Permissible Limit (tons)	Excess (tons)	Permissible Limit (tons)	Excess (tons)
Lemont	18.1	8.90	3.30	5.60	1.60	7.20
Hanover Park	66.5	34.8	12.5	22.3	6.20	28.6
John E. Egan	254	114	36.2	77.5	18.1	95.6
James C. Kirie	206	43.1	50.8	**(7.61)**	25.4	17.8
North Side	1,300	506	380	126	190	316
Calumet	2,570	1,160	361	795	181	976
Stickney	6,550	1,300	1,052	247	526	773
Total	11,000	3,160	1,900	1,270	948	2,210

2.2 BNR Cost Estimates

The costs presented in this report assume the use of the modified five-stage Bardenpho (with methanol addition) technology.[2] The capital costs for upgrading the seven WRPs with the Bardenpho system have been estimated to be $1.51 billion (Table 2-3). This equates to a total present value cost of $2.51 billion.[3]

---

2 Should the MWRDGC be required to meet a new nutrient effluent limit, this particular biological nutrient removal technology may or may not be used; however, the estimated costs reflect what might be required. Still, the nutrient removal cost analysis is hypothetical and only serves to demonstrate the cost differences between wetland and in-plant treatment technologies. The five-stage Bardenpho process is capital cost intensive, requiring extensive tankage, chemicals, and energy. Less capital-intensive options may be applicable for future nutrient standards, and these options would affect the cost comparisons.

3 For present value calculation, interest rate was 4% and payment period was 50 years for the land and 20 years for the facilities.

Table 2-3. Advanced Treatment Capital and Operating and Maintenance (O&M) Costs.

WRP	Advanced Treatment (five-stage Bardenpho Process) (10^3 $)			2002 O&M Costs	Advanced Treatment O&M[2]
	Land	Facilities[1]	Total	(10^3 $)	(10^3 $)
Lemont	158	5,940	6,100	794	397
Hanover Park	636	21,000	21,600	2,900	1,450
John E. Egan	951	40,200	41,200	4,930	2,470
James C. Kirie	924	96,500	97,400	6,780	3,400
North Side	4,080	235,000	239,000	22,100	11,100
Calumet	1,080	250,000	251,000	34,200	17,100
Stickney	3,670	847,000	850,000	78,800	39,000
Total	11,500	1,500,000	1,510,000	151,000	75,000

[1] Includes 15% construction contingency, 15% construction overhead and profit, 15% of total construction costs for engineering, legal, and administration.

[2] 50% of total conventional treatment O&M costs as recommended in Zenz (2003).

2.3 Annualized Treatment Costs

For purposes of developing the cost equations, the annualized capital (facilities and land) and O&M costs for nutrient removal were computed (Table 2-4). The capital costs for six of the seven plants were based on the expansion costs at the Calumet WRP. Internal cost factors for small, medium, and large plants were derived by the MWRDGC and multiplied by the cost of the Calumet WRP; therefore, the listed capital costs do not take into consideration plant specifics, such as influent characteristics that can and will result in design/costs differences. Variations in the specific plants may require more capital upgrade than for others but, on average, the engineering estimates reflect the expected, average cost for the system as a whole.

The MWRDGC researched multiple processes that would treat TN and TP simultaneously or separately to achieve the very low effluent levels of 3.0 mg/L and 0.50 mg/L, respectively. Based on this research, the MWRDGC selected BNR processes and allocated the total annualized costs between TN removal (60.4 %) and TP removal (39.6%) (Table 2-4).

The O&M costs incorporated both the variable and fixed variable costs for labor, energy, chemicals, and supplies. The variable costs are those costs that fluctuate with changes in output; whereas, fixed variable costs are those costs that do not vary with the level of removal but would be eliminated if the process were to be shut down or negated (e.g., labor). The advanced treatment O&M costs were estimated under several different scenarios. Employing an average O&M cost ($0.854/1000 gallons) that incorporated power, chemical, and labor costs for BNR treatment (Reddy, 1998), the total annual O&M costs were from $17,800 to $102,000 per ton of nutrient (TN and TP combined) removed based on the less stringent effluent standards for the seven WRPs. With the minimum cost equation ($0.252/1000 gallons), the O&M costs calculated were from $5,200 to $30,100 per ton of nutrient removed. These costs estimates are for an entire BNR system that includes nitrification. The MWRDGC, based on its operating experience, estimates that Bardenpho O&M costs would add 50% to the current (2002) O&M cost, as reported in Zenz (2003) (Table 2-3). The O&M costs under this scenario were from $4,080 to $15,400 per ton of nutrient removed (both TN and TP) ($0.090–$0.320/1000 gallons). O&M costs of nitrogen removal were from $0.054–$0.200/1000 gallons. It should be noted that other cost estimate studies reported lower O&M costs for nitrogen removal at $0.040–$0.090/1000 gallons (Moore et al., 2000) and $0.040–$0.076/1000 gallons (Bacon & Pearson, 2002).

Table 2-4. Annualized Advanced Treatment Costs.

WRP	Costs (10^3 $)				Allocation ($)	
	Land[1]	Facilities[2]	O&M	Total	TN (60.4%)	TP (39.6%)
Lemont	3.66	432	397	830	502	330
Hanover Park	29.4	1,520	1,450	3,000	1,810	1,190
John E. Egan	44.0	2,920	2,470	5,430	3,280	2,150
James C. Kirie	42.8	7,020	3,400	10,500	6,310	4,140
North Side	189	17,100	11,100	28,300	17,100	11,200
Calumet	50.0	18,200	17,100	35,300	21,300	14,000
Stickney	170	61,600	39,400	101,000	61,000	40,100
Total	529	109,000	75,300	184,000	111,000	73,100

[1] Annualized cost calculation based on n = 50 years, i = 4%.
[2] Annualized cost calculation based on n = 20 years, i = 4%.

2.4 BNR Nutrient Removal Cost Equations

For each plant, the daily advanced treatment cost was determined separately for TN and TP removal using the following equation:

$$\text{Advanced Trt Costs (\$)} = \left(\frac{\text{Annualized Capital Cost (\$)}}{365} \right) + \left[\frac{\text{Daily Excess Tons}}{\text{Annual Excess Tons}} \right] \left(\text{O\&M Costs (\$)} \right) \qquad (2.4\text{-}1)$$

Because the annualized capital costs are independent of nutrient loadings, these costs were allocated evenly over 365 days. In contrast, the daily O&M costs depend on the nutrient loadings as operation parameters may change for low or high nutrient loadings; therefore, the annual O&M costs were allocated based on the fraction of tons removed daily. The daily costs were summed on a monthly basis. The monthly costs were graphed versus the corresponding monthly excess nutrient loadings. The total cost equation is the linear fit to the data: $Y = aX + b$, where Y is the nutrient removal cost expressed in dollars, a is the marginal cost ($/ton), b represents the monthly capital costs ($), and X represents the tons of nutrient (TN or TP) needed to be removed monthly to meet the permissible limit. Examples of the TN and TP cost functions for the Calumet WRP are presented in Figure 2-1.

The individual monthly total cost equations for each WRP are listed in Table 2-5. The intercepts represent monthly capital costs. The slopes, or marginal cost, are a function of nutrient loadings and factors of treatment (e.g., labor, chemicals, and energy). The seven individual WRP cost equations were used to project the costs for the lower effluent limits of 2.18 mg/L TN and 0.50 mg/L TP for each plant. The total annual advanced treatment costs for the two TN and TP effluent limits are listed in Table 2-6 and 2-7, respectively. The average cost is the total removal cost divided by the total tons removed. Because the cost equations are linear, the positive y-intercept ($) of the total cost function ensures falling average costs ($/ton) as nutrient removal loads increase with the possibility of positive economies of scale. Because the average cost is decreasing, the marginal cost ($/ton) is less than the average cost at every level of removal.

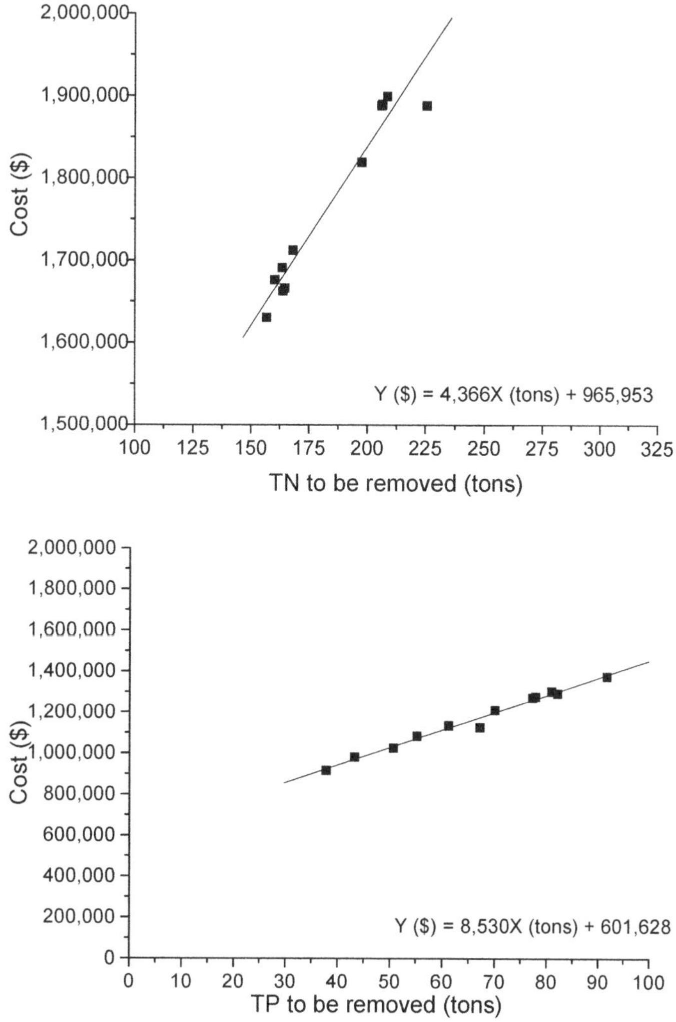

Figure 2-1. Total Cost Curves for the Calumet WRP for TN Removal (top) and TP Removal (bottom).

The daily advanced treatment cost was modified for several of the WRPs that met the higher TP effluent limit. For several days in 2002, the effluent of the Stickney, Kirie, and North Side WRPs met or were below 1.0 mg/L TP. For these dates, mathematically, there were negative loads removed. Therefore, for the days the limit was met, only capital costs were considered in the monthly cost summation, as theoretically there would be no associated O&M costs. Only positive loads removed were used in the development of the cost equation. However, in the application of the cost equations to determine the total advanced treatment cost, the negative loadings were considered because the MWRDGC would save money and could use the extra removal to meet their overall monthly demand. As Kirie had an overall annual TP effluent less than 1 mg/L, the average cost is not applicable and is not given in Table 2-7. In general, the total and marginal costs for TP removal are significantly higher than TN costs because of the higher costs allocated to TP and the smaller mass that is removed.

Table 2-5. Monthly Total Cost Equations for TN and TP Reduction.

WRP	TN Cost Equation[1]				TP Cost Equation[1]			
Lemont	Y =	6,390X	+	18,900	Y =	30,600X	+	13,200
Hanover Park	Y =	10,100X	+	67,800	Y =	27,200X	+	48,700
John E. Egan	Y =	3,130X	+	156,000	Y =	14,300X	+	87,000
James C. Kirie	Y =	10,100X	+	335,000	Y =	152,000X	+	232,000
North Side	Y =	2,810X	+	819,000	Y =	29,300X	+	557,000
Calumet	Y =	4,370X	+	966,000	Y =	8,530X	+	602,000
Stickney	Y =	3,070X	+	3,020,000	Y =	36,600X	+	2,000,000

[1] X = load removed (tons), Y = Cost ($).

Table 2-6. Estimated Annual Costs for TN Reduction.

WRP	Total Annual Cost ($)		Average Cost ($/ton)		Marginal Cost[1] ($/ton)
	3.0 mg/L	2.18 mg/L	3.0 mg/L	2.18 mg/L	
Lemont	502,000	520,000	11,700	11,400	6,390
Hanover Park	1,820,000	1,930,000	18,300	17,600	10,100
John E. Egan	3,280,000	3,380,000	7,300	7,030	3,130
James C. Kirie	6,310,000	6,730,000	27,800	24,900	10,100
North Side	17,100,000	18,000,000	6,610	6,190	2,810
Calumet	21,300,000	22,600,000	9,570	8,930	4,370
Stickney	61,000,000	63,700,000	7,570	7,110	3,070
Total	111,000,000	117,000,000			

[1] Marginal cost independent of effluent limit.

Table 2-7. Estimated Annual Costs for TP Reduction.

WRP	Total Annual Cost ($)		Average Cost ($/ton)		Marginal Cost[1] ($/ton)
	1.0 mg/L	0.5 mg/L	1.0 mg/L	0.5 mg/L	
Lemont	331,000	381,000	59,000	52,900	30,600
Hanover Park	1,190,000	1,360,000	53,400	47,600	27,200
John E. Egan	2,160,000	2,410,000	27,800	25,200	14,300
James C. Kirie[2]	***1,620,000***	5,480,000	**NA**	309,000	152,000
North Side[3]	10,400,000	15,900,000	81,900	50,400	29,300
Calumet	14,000,000	15,500,000	17,600	15.900	8,530
Stickney[3]	33,100,000	52,000,000	134,000	67,700	36,600
Total	63,000,000	93,000,000			

[1] Marginal cost independent of effluent limit.
[2] Kirie met the 1.0 mg/L TP effluent limit for based on monthly totals and had a negative excess load total (Table 2-2).
[3] North Side and Stickney met the 1.0 mg/L TP effluent limit some days in 2002.

CHAPTER 3.0

WETLAND TREATMENT COST ANALYSIS

3.1 Wetland Nutrient Removal Calculations

3.1.1 Nutrient Loads

For purposes of this analysis, a unit treatment wetland was devised, cost parameters were estimated, and cost equations were developed. The unit wetland was assumed to comprise 1500 acres and to be located on the floodplain of the Illinois Water (River), approximately 130 miles downstream of the MWRDGC service area in north-central Illinois. The monthly nutrient concentrations and water temperatures used in the case study were taken from the Illinois River at Ottawa, Ill. (U.S. Geological Survey Station # 05553500). Table 3-1 presents the monthly values. The nutrient and water temperature data for 2 months (September and November) were estimated because the data were not reported. The average of the data for the month before and month after was used to estimate the missing values.

Table 3-1. Monthly Wetland Input Parameters.

Month	Season	Temperature (°C)	Inlet TN Concentration[1] (mg/L)	Inlet Nitrate + Nitrite Concentration[1] (mg/L)	Inlet TP Concentration[1] (mg/L)	k (TP)[2] (m/yr)
Jan	Winter	2.1	6.40	5.00	0.73	7.0
Feb	Winter	3.2	8.50	7.10	0.35	9.0
March	Spring	7.3	7.29	6.29	0.35	7.5
April	Spring	15.8	6.46	5.36	0.37	30.1
May	Spring	22.5	5.10	3.70	0.56	22.7
June	Summer	25.0	8.63	7.33	0.36	19.1
July	Summer	29.2	5.10	4.12	0.54	12.4
Aug	Summer	26.0	4.32	3.33	0.74	13.6
Sept	Autumn	20.9	4.35	3.30	0.79	15.8
Oct	Autumn	15.8	4.37	3.27	0.83	21.6
Nov	Autumn	11.6	5.09	4.18	0.62	21.8
Dec	Winter	7.4	5.81	5.10	0.41	10.2

[1] 2001USGS water quality data for Illinois River at Ottawa, Ill. (USGS Station #05553500).
[2] Rate constant for phosphorus removal (Kadlec, unpublished data.).

3.1.2 Monthly Removal Calculations

A FSW wetland model that assumes wetlands behave like well-mixed units in series was developed to determine the removal of TN and TP. The conceptual configuration for the 1500-acre unit wetland, or nutrient farm, had three compartments perpendicular to flow. Each compartment was divided into two interior cells. Two interior levees formed the three compartments with each interior levee containing two redistribution structures to provide multiple flow paths. Wetlands are more efficient if operating closer to plug flow (i.e., with many tanks in series).

Mesocosm and field investigations support the assumption that nutrient removal is best represented by a first-order relationship (Crumpton et al., 1993; Kadlec & Knight, 1996; Spieles & Mitsch, 2000). Monthly removals of TN and TP were defined by the following general equation,

$$C_e = C_i \left(1 + {k_N}/{(Nq)}\right)^{-N} \qquad (3.1.2\text{-}1)$$

where C_e = outlet concentration (mg/L),
C_i = influent concentration (mg/L),
k_N = first order areal rate constants for N continuously stirred tank reactors (m/yr),
q = hydraulic loading rate (m/yr),
N = number of well-mixed units assumed (in this model, $N = 2$).

Total nitrogen is composed of ammonia (NH_4-N), nitrate plus nitrite (NO_3-N+NO_2-N), and organic forms (N_{org}). However, ammonia is present at only relatively low levels in the Illinois River at the assumed point of treatment, averaging about 1.11 mg/L. Therefore, only organic and oxidized (nitrate + nitrite) forms are considered further. The amount of N_{org} in the Illinois River is also very small, and is assumed to be negligible. The wetland is expected to generate N_{org} at a concentration of about 1.5 mg/L.

The rate constant for nitrate removal was determined by $k = k_{20} * \Theta^{(\text{temperature} - 20)}$; whereas, the rate constants used for phosphorus removal are listed in Table 3-1. After a period of initial adaptation, the rate of phosphorus removal may change. During the adaptation period, the sequestration of phosphorus species is by existing soil. Once the soil becomes saturated, sustainable removal of phosphorus is from particulate settling and accretion in newly formed sediments. The design calculations used in this study pertain to this long-term stable operation and are based on data from 141 systems that all receive concentrations less than 1.0 mg/L TP (Kadlec, unpublished data). Wetlands located in cold climates can consistently retain phosphorus in amounts of 0.1–6 g m^{-2} yr^{-1} (Kadlec & Knight, 1996).

The removal rate is determined from the difference between the influent concentration and the calculated outlet concentration multiplied by the hydraulic loading rate,

$$Removal\ Rate \left(g\ m^{-2}\ yr^{-1}\right) = 3.65 \left(C_i - C_e\right) q \qquad (3.1.2\text{-}2)$$

where q = hydraulic loading rate (cm/day), 3.65 is the unit conversion factor.

For TN the difference between influent and outlet concentrations in the removal rate calculation (Eq. 3.1.2-2) was calculated as the removal of oxidized nitrogen, plus a 1.5 mg/L gain in N_{org},

$$[TN_{IN}] - \left\{ \left([NO3 + NO2]_{IN} - [NO3 + NO2]_{OUT} \right) + [1.5\, N_{ORG}] \right\} = [TN_{REMOVED}] \quad (3.1.2\text{-}3)$$

The total mass removed for TN and TP was determined as,

$$Load\ removal\ (tons) = Removal\ rate\ \left(g\ m^{-2}\ yr^{-1}\right) * EA * 3.714E-04 \quad (3.1.2\text{-}4)$$

where EA = effective area (acres), 3.714E-04 is the unit conversion factor.

The unit design calculations presume that only two-thirds of the area is effective at any one time. This allows for a conservative estimate of nutrient removal and accounts for any animal "damage" to vegetation that is likely to occur. The presence of emergent plants is important as they assist in the settling of suspended sediment and provides the surface area for microbial biofilm growth.

In this study, the excess loads, which need to be removed to meet the criteria for the seven MWRDGC WRPs, and the in-stream concentrations, at the wetland site were used to determine the total wetland area. Because the models used to forecast performance respond to changes in hydraulic loading and temperature on a monthly basis, the required area varied with these parameters. Consequently, at low temperatures, the land requirements were greater and, in fact, in December and January, land requirements were the greatest. Also, the land requirements would have been greater if it had been assumed that the wetland effluent needed to meet the in-stream water quality standards. In this regard, the treatment wetland only was required to remove the requisite load.

3.2 Wetland Cost Estimates

A large database exists on wetland performance in the context of treatment wetlands (Kadlec & Knight, 1996). There is a firm basis for wetland biogeochemistry and treatment wetland design. The wetland cost and nutrient removal parameters used in this case study were determined based on treatment wetland literature, practical experience in the construction and operation of treatment wetlands in the Midwest and Florida, and construction and restoration costs of large-scale wetlands in Illinois. The parameters used in the model are listed and defined in Table 3-2.

The calculated costs were based on land, wetland restoration, hydraulic controls, and O&M. The costs associated with pumping and water distribution have been minimized because the nutrient-rich water is already being discharged to the river system. Inflow and outflow pumps were assumed necessary. The pumps would be located on the external levee, between the river and the wetlands, and would control the hydraulic loading into and through the nutrient farm. The capital cost for large scale pumping stations can account for a third or more of the total project cost (Kadlec, in press); therefore, if gravity flow can be used, actual pumping costs could be less than those assumed in this study. The capital cost for pumps was determined by the equation,

$$Capital\ Costs\ (\$) = 3{,}500X + 75{,}000 \quad (3.2\text{-}1)$$

where X = pumping capacity, cfs.

The annual energy cost for pumping, either at the inlet or outlet, was based on a static head of 10 feet, and an energy price of \$0.183/kWh,

$$Annual\ Energy\ Costs\ (\$) = 1,357.7X \tag{3.2-2}$$

where X = pumping capacity, cfs.

A pump efficiency of 80% was taken into consideration within the wetland model calculations. The pumping capacity for a given land size was assumed to be,

$$Pumping\ Capacity\ (cfs) = 0.06L \tag{3.2-3}$$

where L = land size (acres).

Table 3-2. Wetland Model Parameters and Assumptions for the Wetland Cost Curve Calculations.

Parameter Category	Parameter	Variable	Value	Definition/Equation
Wetland Basic	Depth	h	2.0 ft	
	Nitrogen rate constant (20°C)	k_{20}	35.0 m/yr	(Kadlec & Knight, 1996)
	Temperature factor	θ	1.09	(Kadlec & Knight, 1996)
	Percent vegetated	F	0.67	
	Effective Area	EA	67%	Equation: EA = F * 100 %
	Compartments	C	3	
	Compartment Efficiency	N	2	
	Hydraulic Efficiency	NC	6	NC = N*C*land area
Wetland Construction	Grading Fraction	G	1.00	Fraction of effective area that requires grading
	External Levee Length		0.0017 miles/acre	
	Internal Levee Length		0.00058 miles/acre	
	Redistribution Structures		4	
Pumping	Pumping Stations		2	
	Capital Cost			Equation: Cost = 3500 * pumping capacity + 75,000
	Pumping Capacity			Equation: Capacity (cfs) = 0.06 cfs/acre * area
	Energy Cost		0.183 \$/kWh	
	Annual Energy Cost			Equation: Cost = 1357.7*pumping capacity
	Energy/Pumping Efficiency		80%	
Land Capital	Land		\$2,200/acre	
	Grading		\$500/acre	Based on effective area
	External Levee		\$792,000/mile	
	Internal Levee		\$500,000/mile	
	Roads/Construction		\$6.67/acre	Based on \$10,000/1500 acres
	Structures		\$10,000/each	
Other Costs	Personnel		\$16,000	\$40/h @ 400 h/yr for 1500 acres
	Sampling/Testing		\$ 5,000	Based on 1500 acres
	Materials & Supplies		\$ 5.33/acre	
	Contingency		10%	Of total capital cost (land and pumping)
Nutrient Cost Allocation	% Allocation		91% N, 9% P	Based on an approximate per mass basis

3.3 Wetland Cost Equations

The addition of downstream nonpoint sources to the wetland hydraulic loading may decrease or increase the "evenness" of the wetland influent concentrations. If there were no additional sources, then the variations in point source effluent would affect the load removed and, consequently, the wetland outlet concentrations. The model was used to forecast performance response to changes in hydraulic loading and temperature on a monthly basis. In this study, the excess loads that needed to be removed to meet the total monthly demand of the seven MWRDGC WRPs were used to determine the total wetland area. If the wetland effluent has to meet the in-stream standard, the required wetland area would be larger than stated.

The total wetland area and associated costs were scaled up from the unit size of 1500 acres. The total annualized cost (farm land purchase, capital improvements, and O&M) was allocated evenly over 12 months regardless of loading rate or removal requirements (Table 3-3). A cost equation was constructed for each season (Table 3-4). The cost allocation between nitrogen and phosphorus removal (91% and 9%, respectively) was based on the approximate mass of each nutrient removed within the wetland.

Table 3-3. Annualized Costs for the Model 1,500-Acre Wetland.

Category	Item	Payment Period[1] (years)	Costs[2] ($)
Land	Purchase	50	154,000
	Salvage	50	**(21,600)**
Land Improvements	External Levee	50	104,000
	Internal Levees	20	35,000
	Control Structures	20	3,240
	Roads	20	809
	Grading	50	25,600
Pumping Capital	Pumping Stations	20	63,100
O&M	Labor and Supplies	NA	29,000
	Electricity	NA	122,000
Total			515,000

[1] Interest at 4%.

[2] Includes contingencies, except for land purchase.

The seasonal total cost equations were derived from the linear regression of the annualized monthly costs versus the monthly removal of nutrient (TN or TP) for different sized nutrient farms (Table 3-4). Graphical representations of the four seasonal curves for TN and TP removal are presented in Figure 3-1.

Table 3-4. Total Cost Equations for TN and TP Removal in Wetlands.

Season	TN Cost Equation[1]		TP Cost Equation[1]	
Winter	Y = 2,740X + 650,000	R^2 = 0.855	Y = 2,350X + 96,000	R^2 = 0.784
Spring	Y = 1,990X + 239,000	R^2 = 0.946	Y = 2,610X + 99,000	R^2 = 0.776
Summer	Y = 960X + 1,700,000	R^2 = 0.619	Y = 1,590X + 123,000	R^2 = 0.722
Autumn	Y = 3,000X + 108,000	R^2 = 0.976	Y = 1,400X + 17,500	R^2 = 0.960

1. X = load removed (tons), Y = Cost ($).

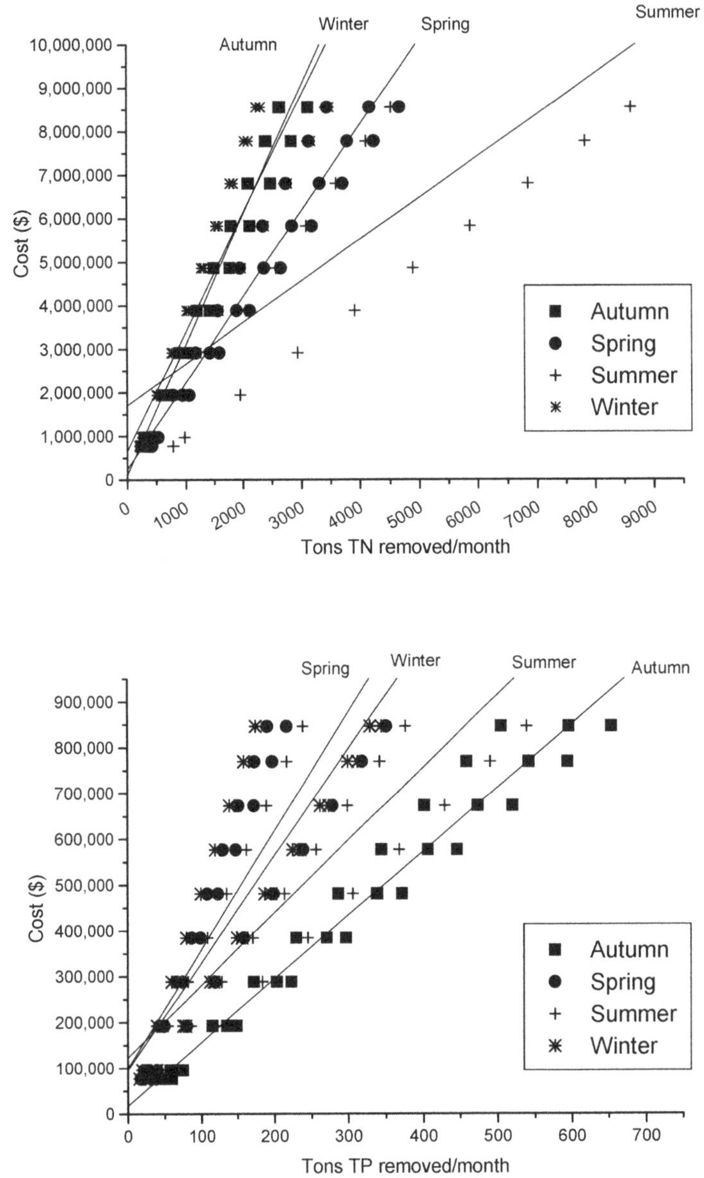

Figure 3-1. Total Cost Curves for the Treatment Wetland for TN Removal (top) and TP Removal (bottom).

In this case study, the wetlands were operated at a constant hydraulic loading throughout the year; however, during periods when the concentrations of nitrate are high in the source water (i.e., spring) the hydraulic loading could be increased to increase the mass of NO_3-N removed. The overall concentration reduction would be less, as there is a trade-off between removal efficiency and load reduction. Under the parameters used in this model, the highest removal efficiency is in summer for TN (77%) and spring and autumn (58%) for TP (Table 3-5). The lowest removal efficiencies were in January for both TN (27%) and TP (33%), hence January is the critical month in determining the wetland area needed to meet the monthly demand of the MWRDGC.

Table 3-5. Nutrient Loads Removed and Removal Efficiency for Model Wetland.

Month	TN				TP		
	Hydraulic Loading (cm/d)	Outlet Concentration (mg/L)	Loads Removed (tons)	Removal Efficiency (%)	Outlet Concentration (mg/L)	Loads Removed (tons)	Removal Efficiency (%)
Jan.[1]	4.4	4.7	10.17	27	0.48	1.50	35
Feb.	4.4	5.8	15.81	32	0.22	0.79	38
March	4.4	4.6	15.65	36	0.20	0.87	42
April	4.4	2.9	21.20	56	0.20	0.99	45
May	4.4	1.9	18.95	63	0.29	1.59	48
June	4.4	2.0	39.13	77	0.18	1.08	51
July	4.4	1.6	20.56	68	0.25	1.71	54
Aug.	4.4	1.7	15.54	61	0.32	2.45	56
Sept.	4.4	1.9	14.17	55	0.33	2.71	58
Oct.	4.4	2.3	12.01	47	0.33	2.97	61
Nov.	4.4	3.1	11.96	40	0.23	2.29	63
Dec.	4.4	4.0	10.51	31	0.15	1.56	65

Because nitrate reduction by denitrification is microbially mediated, nitrate removal is highest in the summer and spring months in comparison to the colder seasons. In addition to denitrification, nitrogen is removed by plants. The plants use ammonium for growth during the spring and summer months, and have a secondary reliance on NO_3-N. Because the influx of NO_3-N into the Illinois River is seasonal, peaking in the spring, the maximum removal is necessary during this time. Typically, phosphorus removal is highest in the spring as phosphorus is used in new plant growth and lowest during the fall as biomass decomposition returns phosphorus to the system. In this particular case study, the removal of phosphorus was highest in the fall because of the higher incoming water concentrations for the months of September, October, and December. In general, the removal of nitrogen or phosphorus is greater with higher incoming concentrations. The effluent concentrations in this study were 0.15–0.48 mg/L (TP) and 1.6–5.8 mg/L (TN).

The cost of treatment wetlands depends on the amount and complexity of pumping, distribution, control structures, site preparation, and land costs. The costs are much lower for a wetland within the levee district on the Illinois River as only minimal site grading would be necessary. The location, natural topography and hydrology, existing seed bank for vegetation re-establishment, and existence of an external levee will help to alleviate these costs. The straight cost for the unit treatment wetland is approximately $4,860 per acre. Mitsch and Gosselink (2000) placed the cost of a 1500-acre wetland at $3,000 per acre and a 3000-acre wetland at $2,100/acre. Based on the itemized analyses in this case study, these published estimates are relatively low.

CHAPTER 4.0

COST COMPARISONS BETWEEN TRADITIONAL AND WETLAND NUTRIENT MANAGEMENT

4.1 Monthly Cost Comparison

Using the cost equations, seasonal costs tables were constructed (Tables 4-1 and 4-2). These costs reflect the influent and effluent characteristics of the seven MWRDGC wastewater treatment plants, the characteristics of the nutrient load in the Illinois River, and the seasonal performance of the treatment wetlands.

As described above, the performance of the treatment wetlands was seasonally derived; consequently, the required land area and costs were calculated for each month in the season. Based on the less stringent standards (3.0 mg/L TN and 1.0 mg/L TP), the wetlands removed, on an annual basis, 25,900 tons of nitrogen at a cost of $58,200,000 (Table 4-1). On the other hand, the WWTPs only produced 13,700 tons of nitrogen above the standard. The cost to remove precisely this amount via wetlands would be $37,600,000. A similar situation exists for TP (Table 4-2).

Table 4-1. Monthly Comparison of Excess TN Removal and Costs (10^3 $) for 189,000 Acres, which Meets the Limit for 3.0 mg/L TN and 1.0 mg/L TP (Controlling Parameter).

Month	Season	TN				TN Extra Credits	
		Wetland Removal		WRP Requirement			
		Tons	Cost (10^3 $)	Tons	Wetland Cost (10^3 $)	Tons	Value (10^3 $)
Jan[1]	Winter	1,280	4,160	1,190	3,920	90	247
Feb	Winter	1,990	6,110	1,190	3,910	810	2,210
March	Spring	1,970	4,160	1,460	3,130	510	1,020
April	Spring	2,670	5,540	1,220	2,670	1,450	2,870
May	Spring	2,390	4,980	1,150	2,520	1,240	2,460
June	Summer	4,930	6,450	1,100	2,760	3,830	3,690
July	Summer	2,590	4,200	1,100	2,770	1,490	1,430
Aug	Summer	1,960	3,590	1,140	2,810	810	784
Sept	Autumn	1,790	5,460	1,030	3,200	750	2,260
Oct	Autumn	1,510	4,640	1,050	3,260	460	1,380
Nov	Autumn	1,510	4,620	955	2,970	550	1,650
Dec	Winter	1,320	4,280	1,100	3,670	220	614
Total		25,900	58,200	13,700	37,600	12,220	20,600

[1] Critical month in determining the wetland area needed to meet the monthly removal requirement for the controlling parameter (TP).

Table 4-2. Monthly Comparison of Excess TP Removal and Costs (10^3 $) for 189,000 Acres, which Meets the Limit for 3.0 mg/L TN and 1.0 mg/L TP (Controlling Parameter).

Month	Season	TP				TP Extra Credits	
		Wetland Removal		WRP Requirement			
		Tons	Cost (10^3 $)	Tons	Wetland Cost (10^3 $)	Tons	Value (10^3 $)
Jan[1]	Winter	189	540	188	538	0	0
Feb	Winter	100	331	97	323	3	8
March	Spring	109	384	61	258	50	126
April	Spring	124	423	6	115	120	308
May	Spring	201	623	18	145	180	477
June	Summer	136	340	76	244	60	96
July	Summer	215	466	136	339	80	127
Aug	Summer	309	615	122	317	190	298
Sept	Autumn	341	494	139	211	200	282
Oct	Autumn	374	540	162	243	210	296
Nov	Autumn	289	421	148	224	140	197
Dec	Winter	197	560	114	365	80	195
Total		2,580	5,740	1,270	3,320	1,320	2,410

[1] Critical month in determining the wetland area needed to meet the monthly removal requirement for the controlling parameter (TP).

Because the required land area was based on meeting the total monthly demand during the coldest month of the year, the annual masses of TP and TN removed were well beyond the MWRDGC's need. For example, 189,000 acres of wetland are required to meet both the 3.0 mg/L TN and 1.0 mg/L TP standard, so that the demand of the critical month, January, is met; in most cases, phosphorus removal is the controlling parameter. If treatment wetlands of this area were operated, they would remove 12,200 extra tons of TN (25,900 tons removal minus the WRP requirement of 13,700 tons). The extra tons and costs for each standard are summarized in Table 4-3.

The nutrient farm operator could reduce pumping during the noncritical periods, thereby reducing his operating costs. Alternatively, the farmer could continue to remove the excess nutrients and sell the surplus to other WWTPs or to other sources that need to reduce their emissions, such as agriculture, energy producers and manufacturing.

Table 4-3. Summary Table of Nutrient Removal (tons) and Cost (10^3 $) for Wetlands and Conventional Treatment.

Nutrient Limit (mg/L)	Land Size (acres)	Wetland Removal		WRP Requirement		Extra Tons and Credits	
		Tons	Cost (10^3 $)	Tons	Cost (10^3 $)	Tons	Cost (10^3 $)
3.0 TN	176,000	24,100	54,800	13,700	37,600	10,400	17,200
2.18 TN	193,000	26,500	59,300	15,300	41,000	11,100	18,300
1.0 TP	189,000	2,580	5,740	1,270	3,320	1,320	2,410
0.5 TP	322,000	4,400	9,100	2,210	5,250	2,190	3,810

The total costs and present values for WRPs and wetlands are presented in Table 4-4. To achieve the two sets of effluent limits (3.0 mg/L TN and 1.0 mg/L TP, 2.18 mg/L TN and 0.5 mg/L TP), 189,000 and 322,000 total acres of wetland are needed. Even with this land requirement, employing the use of wetlands instead of advanced treatment processes to meet regulatory nutrient loadings would result in an annual savings from 63% to 51%, depending on the standard. There is fewer savings for the more stringent standards because more land is required to achieve the 0.50 mg/L TP standard. If the extra credits produced in the wetlands were sold at cost, the savings would increase to 76% and 78%, depending on the standard (Table 4-5). In this case, there is a greater savings at the lower standard limits because of the increase in the number of available nitrogen credits for the larger wetland size requirement.

Table 4-4. Total Annual and Present Value Cost Comparison (10^3 $).

Criteria Limit	Total Cost (10^3 $)			Present Value[3] (10^3 $)			
	Wetland Total[1]	WRP Total[2]	Savings	Wetland Total	WRP Total	Savings	%
3.0 mg/L TN 1.0 mg/L TP	63,900	174,000	110,000	870,000	2,370,000	1,500,000	63
2.18 mg/L TN 0.50 mg/L TP	103,000	211,000	108,000	1,390,000	2,860,000	1,470,000	51

[1] For land purchasing costs and salvage value *n* = 50 years, *i* = 4%; land improvements (grading, levee construction, roads) and pumping capital: *n* = 20 years, *i* = 4%.
[2] For land costs: *n* = 50 years, i = 4%; capital costs *n* = 20 years, *i* = 4%.
[3] *n* = 20 years, *i* = 4%.

Table 4-5. Total Annual and Present Value Cost Comparison, Including Sale of Extra Credits (10^3 $).

Criteria Limit	Total Cost (10^3 $)			Present Value[3] (10^3 $)			
	Wetland Total[1]	WRP Total[2]	Savings	Wetland Total[1]	WRP Total	Savings	%
3.0 mg/L TN 1.0 mg/L TP	40,900	174,000	133,000	556,000	2,370,000	1,810,000	76
2.18 mg/L TN 0.50 mg/L TP	46,000	211,000	164,000	629,000	2,860,000	2,230,000	78

[1] For land purchasing costs and salvage value *n* = 50 years, *i* = 4%; land improvements (grading, levee construction, roads) and pumping capital *n* = 20 years, *i* = 4%.
[2] For land costs: *n* = 50 years, *i* = 4%; capital costs: *n* = 20 years, i = 4%.
[3] *n* = 20 years, *i* = 4%.

4.2 Marginal Cost Comparison

The marginal costs for nutrient removal are given in Table 4-6. The marginal cost represents the price paid, or cost incurred, to remove the next ton or the last ton of nutrient. Because the cost functions were found to be linear, the marginal costs of the WRPs were the same throughout the operating range. On the other hand, marginal costs of wetlands vary from season to season but the intra-seasonal marginal costs were found to be liner. For example, the marginal cost for removing a ton of nitrogen in the winter would be $2,740, whereas in the summer it would be $960. The marginal cost for wetlands was $960–$3,000. This range, of course, depended on temperature as well as the influent nutrient concentrations to the wetland. The lower the concentration, the higher the marginal cost of treatment. Similarly, the lower the water temperature, the higher the marginal cost. Marginal cost at the WRPs did not vary seasonally, for the influent concentrations, temperatures, and removal efficiencies were reasonably uniform. Still, the marginal cost among the plants varied greatly, particularly for phosphorus. The marginal costs for nitrogen removal was from $2,810–$10,100/ton. On the other hand, marginal costs for phosphorus were \$8,500–\$152,000/ton. The high end of the range results from the efficiency with which the Kirie WRP is now operated. On many days, this plant met the 0.5 mg/L TP effluent limit and, of course, satisfied the less stringent effluent limit a greater percent of the time.

Table 4-6. Comparison of Marginal Cost (MC) for TN and TP: Wetland and Conventional Treatment.

Conventional Treatment (\$/ton)			Wetland (\$/ton)		
WRP	**TN**	**TP**	**Season**	**TN**	**TP**
Lemont	6,400	30,600	Winter	2,740	2,350
Hanover Park	10,100	27,200			
John E. Egan	3,130	14,300	Spring	1,990	2,610
James C. Kirie	10,100	152,000			
North Side	2,810	29,300	Summer	960	1,590
Calumet	4,370	8,500			
Stickney	3,070	36,600	Autumn	3,000	1,400
Load-Weighted MC	3,410	16,000	Load-Weighted MC	1,930	1,830

4.3 Comparison to Long Island Sound Estimates

The cost functions and estimates for nutrient removal by conventional treatment technology compare favorably with the removal costs derived from the nitrogen credit trading program established in Connecticut for Long Island Sound (LIS) (Moore et al., 2000; Rocque, 2003). Considering the 2002 costs for nitrogen removal at seven wastewater treatment plants tributary to the LIS, which are similar in actual flow to the MWRDGC WRPs, the load-weighted average cost per ton for nitrogen removal was \$6,870 (Table 4-7). This is somewhat less than the load-weighted average cost of \$8,130 for the MWRDGC's seven plants.

Table 4-7. Long Island Sound Nitrogen Removal Costs.

Connecticut WWTPs	1990 Actual Flow[1] (MGD)	Load Removed[2] (tons/yr)	Annual Costs[3] (\$)	Average Cost (\$/ton)
Fairfield	7.57	69.0	1,400,000	20,400
Greenwich	9.82	164.8	187,000	1,140
New Haven East	38.72	528.2	1,260,000	2,390
Stamford[4]	15.83	240.9	4,780,000	19,800
UConn	3.00	7.7	118,000	15,400
Waterbury	22.69	362.8	1,690,000	4,650
West Haven	7.82	31.2	221,000	7.080
			Load-Weighted	6,870

[1] Moore et al., 2000.
[2] 2002 annual E-pounds removed converted to end-of-pipe load by dividing by E-factor (trading ratio).
[3] Annual cost = TN portion capital cost (assuming i = 4%, n = 20 years) + O&M cost.
[4] Includes projected capital cost, O&M, and load removed for facilities to be completed in 2005.

There are several factors that should be considered in this average cost comparison. The average costs for the MWRDGC's plants reflect an effluent standard of 3 mg/L; whereas the LIS costs are based on the load reductions assigned to individual plants through a wasteload allocation program. The effluent TN concentrations for these plants are most likely greater than 3 mg/L. Second, capital cost for the LIS program are significantly less than MWRDGC's plant costs because the most cost-effective projects are undertaken and the selected technology does not need to meet the same stringent standard as the MWRDGC. Those LIS facilities that decide not to implement additional treatment controls can purchase nitrogen credits from facilities that remove more than their allocation requires. Finally, the annual load removed for the Connecticut WWTPs was calculated as the difference between the 2002 performance for the facilities and the baseline estimated in the LIS TMDL. Even when considering the lower average cost, the treatment wetlands in this case study are more economical.

CHAPTER 5.0

SUPPLY AND DEMAND

The foregoing discussion, comparing the cost nutrient removal at WRPs and in wetlands, assumed MWRDGC had purchased land and operated the wetlands. However, if the MWRDGC chose to rely on others to own and operate the wetlands, it could achieve nutrient removal through a contractual relationship with the wetland owner/operator ("nutrient farmer"). As such, the MWRDGC would purchase nutrient removal credits via some market mechanism.

5.1 Demand of WWTPs

According to microeconomic theory, as the price of nutrient credits rises, the demand for credits diminishes. Then, the MWRDGC would buy nutrient credits at any price less than the cost of producing those credits themselves. The demand curve for each nutrient was constructed by, first, determining 12 monthly removal loads for each WRP. The loads removed ranged from the minimum monthly load required to satisfy the less strict criterion to the maximum monthly load required to satisfy the stricter criterion. The tons to be removed were calculated by subtracting the 12 individual removal loads from the maximum monthly load plus one ton, which was added to avoid a zero "tons to be removed" value.

The removal cost for the tons removed was calculated by multiplying the marginal cost by the associated tons removed. These four variable sets (marginal cost, tons removed, tons to be removed, and removal cost) were ranked in ascending order according to their marginal cost and then separated into 12 classes of seven. The load-weighted marginal cost (i.e., weighted price) for each class was calculated by dividing the sum removal costs by the total tons of nutrient removed in that class. In addition, for each class, the average tons to be removed, or delta, was calculated by dividing the total tons to be removed by seven. For each of the seven classes, the aggregate load removed was calculated. The first class aggregate load was the sum of the individual class deltas. For each succeeding class, the aggregate load was determined by subtracting the delta of the preceding class from the preceding sum.

To construct the demand curves, the 12 weighted prices were plotted against the corresponding aggregate load. The resulting curves, shown in Figure 5-1, give the amount of removal credits that the MWRDGC would be willing to buy if the price were less than the cost of removal. For example, the MWRDGC should be willing to purchase 250 tons of nitrogen removal credits if the price is less than $3,000/ton.

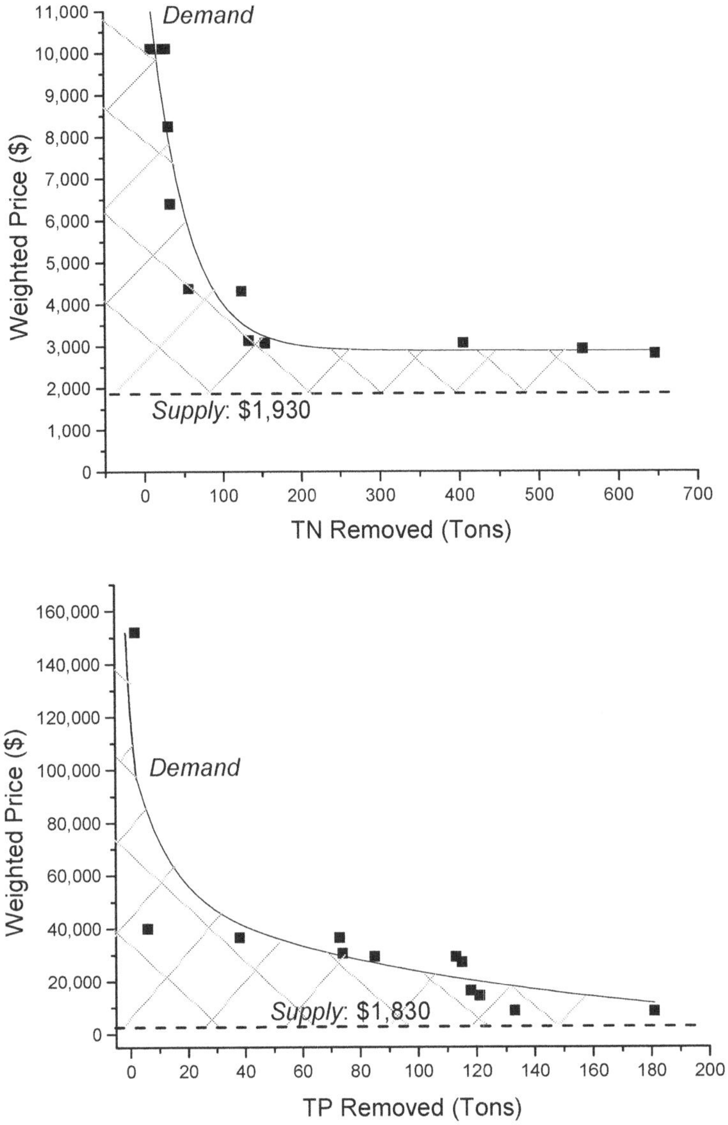

Figure 5-1. Demand and Supply Curves for the Monthly TN Removal (top) and TP Removal (bottom).

5.2 Nutrient Farm Supply

The supply curve was constructed differently. Assuming the availability of an infinite amount of suitable land at a fixed price, as demand increases, the supply would increase simply by adding more wetlands. These treatment wetlands would be operated under the same cost conditions as the other wetlands were operated. The fixed supply price was calculated as the load-weighted seasonal marginal cost. The supply curve, under these conditions, is a horizontal line spanning the range of demand. From high to low prices, the demand curve slopes towards the supply curve but the two curves do not intersect. Consequently, over the range of analysis, there is not an equilibrium, or an intersecting, point beyond which the MWRDGC might well produce its own nitrogen removal credits. Over the range of consideration, it is cheaper for the

MWRDGC to buy the wetland removal credits. The only factor that might alter this conclusion would be that of profit but, for now, profit is not considered within the supply curve.

The shaded area between the supply and demand curves represents the cost savings to the MWRDGC or the profit to the nutrient farmer. How the profits/savings are divided depends on negotiations between the two parties or as stipulated in their contractual agreement. As illustrated in Table 5-1, the cost of nutrient removal per ton for the wetlands is significantly less than the WRP treatment processes. This average cost (total cost divided by total tons of nutrient removed) differential leads to a region of substantial cost savings or profits. In the case where the MWRDGC need to meet the less stringent TN criterion, a total savings of $73,800,000 a year would be what is at stake (Table 5-2). If this sum were split evenly, the nutrient farmer might walk away with $36,900,000 in net profit. This would equate to $195 of net profit per acre, a very handsome sum for any farmer in today's market. A similar situation exists for TP. The net profit per acre could be $157 per acre. The aggregate net profit would be $352 per acre. The situation changes a bit if the more stringent criteria must be met. As shown in Table 16, the net total profit would be only $255 per acre, but this is still a handsome profit. Both of these net profits are substantially above what is earned today in the corn and soybean production areas of the Midwest. In Illinois, tillable land rents for about $150 per acre, which is a reflection of the net profit for agricultural production. These savings and profits would provide the economic incentive for bottomland property owners to convert to nutrient farming.

Table 5-1. Average Cost (Load-Weighted) for WRP and Wetland Treatment.

Nutrient	Average Cost ($/ton)	
	WRP	Wetland
TN (3.0 mg/L)	8,130	2,250
TN (2.18 mg/L)	7,660	2,120
TP (1.0 mg/L)	49,500	2,220
TP (0.5 mg/L)	42,200	2,060

Table 5-2. Profit Comparison (Annual Cost Savings Includes Sale of Extra Credits).

Wetland Size (acres)	TN			TP		
	Savings	50% split of savings	Net profit/acre	Savings	50% split of savings	Net profit/acre
189,000	73,800,000	36,900,000	195	59,400,000	29,700,000	157
322,000	76,000,000	38,000,000	118	88,400,000	44,200,000	137

CHAPTER 6.0

Conclusions and Recommendations

6.1 Conclusions

On the basis of the foregoing analyses, treatment wetlands can efficiently remove significant loads of nitrogen and phosphorus. Correctly designed and operated, these wetlands could remove nitrogen to a level envisioned in the nutrient criteria established by the U.S. EPA in 2001. In addition, the wetlands could remove a substantial amount of phosphorus—though perhaps not as much as the U.S. EPA envisions. The cost of restoring and operating treatment wetlands is 50–60% less than the cost of constructing and operating advanced wastewater treatment systems. The savings could be even greater if secondary markets for the surplus nitrogen and phosphorus credits could be developed. The savings could reach as high as 60–70% of the cost of conventional nutrient control. The surplus credits, generated mostly during the summer months could be sold to other point (e.g., power utilities, automobile owners, or manufacturers) or NPS emitters.

Besides the savings in capital and operating costs, treatment wetlands would provide benefits to the overall ecological health of the river system. Deleterious algal blooms in fresh water ecosystems could be greatly reduced if sufficient phosphorus were removed by the wetlands; public health would be protected by the reduction of NO_3-N in surface waters, which are used for human consumption. Nutrient farms also would support a wide range of aquatic and terrestrial wildlife, including waterfowl and fish. Consequently, recreational opportunities could be developed and recreational income generated.

The land requirements for nutrient farming, although extensive, can be secured. Using the watershed of the case study, the Illinois River contains 1,400,000 acres of flood prone land as defined by the Federal Emergency Management Agency. These are lands that should not be used for agricultural production, let alone for residential, commercial or industrial use. Converting these lands to nutrient farming would increase flood storage and reduce flood damage. These lands, with their hydric soils, would make ideal nutrient farms.

If nutrient farms, or treatment wetlands, are used to meet the needs of municipal and industrial point source dischargers, there will need to be a financial structure to support these landscapes. A range of ideas is possible. One model would be for point source dischargers to acquire land and operate their own nutrient farms. This would link the operation of the nutrient farm closely to the needs of the nitrogen emitters. Another model would be to leave the land in private ownership, perhaps even in the current ownership, and rely upon a contract to govern the relationship between the nutrient emitter and the nutrient farmer. Among other things, such a contract could call for removal to be synchronized with emissions and allow for the nutrient farmer to purchase extra credits on the spot market during critical cold weather periods. At any rate, the nutrient farmer would have the assurance of a long-term period during which revenue would flow into the nutrient farm on a reasonably secure basis and the emitter would have a reasonably long-term, inexpensive solution for nutrient control. A more sophisticated and elaborate method would be to establish nutrient credit markets where nutrient farmers would offer to harvest nutrients at some specific time in the future. Such contracts would be shorter term but, perhaps, less expensive. Whatever the model might be it will need to be tested thoroughly before implementation.

6.2 Recommendations

Although the authors feel confident in the validity of the cost functions and the subsequent analysis and comparison of the alternative treatment technologies, the various assumptions and, ultimately, the costs associated with constructing and operating large-scale nutrient farms need to be proven. In this regard, pilot projects should be established in various ecoregions to study the design, operation, and economic efficiency of treatment wetlands as nutrient farms. Because WWTPs are responsible to state agencies to meet water quality standards, the reliability of nutrient farming must be established.

Pilot projects are needed to compile information on construction and operating costs, maximum nutrient loading rates, scaling-up considerations, seasonal variations in nutrient removal, flood storage, and wildlife effects. In addition, special consideration should be given to quantifying and monetizing the ancillary benefits of sediment and carbon sequestration in treatment wetlands. Another consideration is the energy savings, and thereby the reduction of carbon emissions, of conventional WWTPs. Until large-scale nutrient farms exist, and greater experience gained in their design and operation, there will continue to be reluctance to use this technology on a broad scale.

If pilot projects bear out the results of this study, water quality will be greatly improved and municipal and industrial wastewater treatment will save an enormous amount of money and energy (reducing carbon emissions by implication). At the same time, nutrient farms will spur recreational development and reduce flood damage. Combined, these economic and social benefits should far exceed the cost of shifting from corn and bean production and the cost of constructing and operating nutrient farms.

REFERENCES

Bacon, L.; Pearson, C.N. *Nitrogen Credit Trading in Maryland: A Market Analysis for Establishing a Statewide Framework,* WERF Report 97-IRM-5E; Water Environment Federation: Alexandria, VA, 2002.

Bowden, W.B. The Biogeochemistry of Nitrogen in Freshwater Wetlands, *Biogeochemistry* **1987**, *4,* 313–348.

Brunner, C.W.; Smith, J.W.; Stucky, N.P.; Kempf, D.W.; Baskett, R.K.; Kadlec, R.H.; Patterson, L.B.; White, J.E.; Zdeb, J.A. Design of Constructed Wetlands with Effluent Reuse Wetlands in Columbia, Missouri. In *Proceedings of the Water Environment Federation 65th Annual Conference;* Water Environment Federation: Alexandria, VA, 1992.

Cooke, J.G. Phosphorus Removal Processes in a Wetland after a Decade of Receiving a Sewage Effluent. *J. Environ. Qual.* **1992**, *21(4),* 733–739.

Crumpton, W.G. Using Wetlands for Water Quality Improvement in Agricultural Watersheds; The Importance of a Watershed Scale Approach. *Water Sci. Technol.* **2001**, *44(11-12)*, 559–564.

Crumpton, W.G.; Fisher, S.W.; Isenhart, T.M.; Matter, M.P.; van der Valk, A.G. *Transformations and Fate of Nitrate and Atrazine in Freshwater Wetlands*, Project Completion Report; Leopold Center for Sustainable Agriculture: Ames, Iowa, 1994.

Crumpton, W.G.; Isenhart, T.M.; Fisher, S.W. The Fate of Nonpoint Source Nitrate Loads in Freshwater Wetlands: Results from Experimental Wetland Mesocosms. In *Constructed Wetlands for Water Quality Improvement*; Morshiri, G.A., Ed.; Lewis Publishers: Boca Raton, Fla.,1993.

Dahl, T.E. *Wetland Losses in the United States: 1780s to 1980s;* U.S. Department of Interior: Washington, DC, 1990.

Dahl, T.E. *Status and Trends of Wetlands in the Conterminous United States 1986 to 1997;* U.S. Department of the Interior, Fish and Wildlife Service: Washington, DC, 2000.

David, M.D.; Gentry, L.E. Anthropogenic Inputs of Nitrogen and Phosphorus and Riverine Export for Illinois, USA. *J. Environ. Qual.* **2000**, *29,* 494–508.

Goolsby, D.A.; Battaglin, W.A.; Lawrence, G.B.; Artz, R.S.; Aulenbach, B.T.; Hooper, R.P.; Keeney, D.R.; Stensland, G.J. *Flux and Sources of Nutrients in the Mississippi-Atchafalaya River Basin: Topic 3 Report for the Integrated Assessment of Hypoxia in the Gulf of Mexico.* National Oceanic and Atmospheric Administration, Coastal Ocean Program: Washington, DC, 1999.

Hey, D.L. Nitrogen Farming: Harvesting a Different Crop. *Restor. Ecol.* **2002**, *10(1),* 1–10.

Hey, D.L.; Kenimer, A.L.; Barnett, K.R. 1994. Water-Quality Improvements by 4 Experimental Wetlands. *Ecol. Eng.* **1994**, *3(4),* 384–397.

Johnston, C.A. Sediment and Nutrient Retention by Freshwater Wetlands: Effects on Surface Water Quality. *Crit. Rev. Environ. Contr.* **1991**, *21*, 491–565.

Kadlec, R.H. Pond and Wetland Treatment. *Water Sci. Technol.* **2003**, *48(5),* 1–8.

Kadlec, R. H. "Constructed Wetlands to Remove Nitrate. In *Nutrient Management in Agricultural Watersheds – A Wetlands Solution*; E.J. Dunne, K.R. Reddy, K.R., Carton, O.T., Eds.; Wageningen Academic Publishers: Wageningen, Netherlands, in press.

Kadlec, R.H.; Hammer, D.E. Modeling Nutrient Behavior in Wetlands. *Ecol. Model.* **1988**, *40*, 37–66.

Kadlec, R.H.; Knight, R.L. *Treatment Wetlands*; CRC Lewis Publishers: New York, 1996.

Keenan, L.W.; Lowe, E.F. Determining Ecologically Acceptable Nutrient Loads to Natural Wetlands for Water Quality Improvements. *Water Sci. Technol.* **2001**, *44(11–12),* 289–294.

Kim, S-Y.; Geary, P.M. The Impact of Biomass Harvesting on Phosphorus Uptake by Wetland Plants. *Water Sci. Technol.* **2001**, *44(11–12),* 61–67.

Lund, L.J.; Horne, A.J.; Williams, A.E. Estimating Denitrification in a Large Constructed Wetland using Stable Nitrogen Isotope Ratios. *Ecol. Eng.* **2000**, *14(1–2),* 67–76.

Mitsch, W.J.; Gooselink, J.G. *Wetlands;* Van Nostrand Reinhold: New York, 1993.

Mitsch, W.J.; Gooselink, J.G. *Wetlands,* 3rd Edition; John Wiley and Sons, Inc.: New York, 2000.

Mitsch, W.J.; Day, J.W.; Gilliam, J.W.; Groffman, P.M.; Hey, D.L.; Randall, G.W.; Wang, N. *Reducing Nutrient Loads, Especially Nitrate-Nitrogen, to Surface Water, Ground Water, and the Gulf of Mexico: Topic 5 Report for the Integrated Assessment of Hypoxia in the Gulf of Mexico.* National Oceanic and Atmospheric Administration, Coastal Ocean Program: Washington, DC, 1999.

Moore, R.E.; Overton, M.S.; Norwood, R.J.; DeRose, D.; Corbin, P.G. *Nitrogen Credit Trading in the Long Island Sound Watershed*; Water Environment Federation: Alexandria, VA, 2000.

Moraghan, J.T. Loss and Assimilation of ^{15}N-Nitrate Added to a North Dakota Cattail Marsh. *Aquat. Bot.* **1993**, *46(3-4),* 225–234.

Murkin, H.R. The Basis for Food Chains in Prairie Wetlands. In *Northern Prairie Wetlands*; van der Valk, A.G.; Ed.; Iowa State University Press: Ames, Iowa, 1989.

Nungesser, M.K.; Chimney, M.J. Evaluation of Phosphorus Retention in a South Florida Treatment Wetland. *Water Sci. Technol.* **2001**, *44(11–12),* 109–115.

Phipps, R.G. *Nitrate Removal Capacity of Wetlands.* Ph.D. Dissertation, Iowa State University, Ames, Iowa, 1997.

Rabalais, N.N. Hypoxia in the Gulf of Mexico. In *CoastView Vol. 2: Northern Gulf of Mexico Imagery and Information;* National Oceanic and Atmospheric Administration: Washington, DC, 2004. Available at http://www.csc.noaa.gov/products/gulfmex/html/rabalais.htm.

Rabalais, N.N.; Turner, R.E.; Justic, D.; Dortch, Q.; Wiseman, W.J. *Characterization of Hypoxia: Topic 1 Report for the Integrated Assessment of Hypoxia in the Gulf of Mexico.*

National Oceanic and Atmospheric Administration, Coastal Ocean Program: Washington, DC, 1999.

Reardon, R. *Cost Implication of Design for Nutrient Removal*. Presentation at the 67th Annual Water Environment Federation Technical Exposition Conference, Chicago, 1994.

Reddy, G.B.; Hunt, P.G.; Phillips, R.; Stone, K.; Grubbs, A. Treatment of Swine Wastewater in Marsh-Pond-Marsh Constructed Wetlands. *Water Sci. Technol.* **1991**, *44(11–12),* 545–550.

Reddy, K.R.; Patrick, W.H. Nitrogen Transformation and Loss in Flooded Soils and Sediments. *Crit. Rev. Environ. Contr.* **1984**, *13(4),* 273–309.

Reddy, M. *Biological and Chemical Systems for Nutrient Removal*, Water Environment Federation: Alexandria, VA, 1997.

Reilly, J.F.; Horne, A.J.; Miller, C.D. Nitrate Removal from a Drinking Water Supply with Large Free-Surface Constructed Wetlands prior to Groundwater Recharge. *Ecol. Eng.* **2000**, *14(1-2)*, 33–47.

Richardson, C.J. Mechanisms Controlling Phosphorus Retention Capacity in Wetlands. *Science* **1985,** *228*,1424–1427.

Richardson, C.J.; Craft, C.B. Effective Phosphorus Retention in Wetlands: Fact or Fiction? In *Constructed Wetlands for Water Quality Improvement*; Moshiri, G.A., Ed.; Lewis Publishers: Boca Raton, Fla., 1993.

Richardson, C.J.; Walbridge, M.R.; Burns, A. *Soil Chemistry and Phosphorus Retention Capacity of North Carolina Coastal Plain Swamps Receiving Sewage Effluent*, Report 212; North Carolina Water Resources Research Institute: Raleigh, NC, 1988.

Rocque, A.J. *Report of the Nitrogen Credit Advisory Board*; Connecticut Department of Environmental Protection: Hartford, Conn., 2003.

Spieles, D.J., Mitsch, W.J. The Effects of Season and Hydrologic and Chemical Loading on Nitrate Retention in Constructed Wetlands: A Comparison of Low- and High-Nutrient Riverine Systems. *Ecol. Eng.* **2000**, *14* (1–2): 77–91.

U.S. Environmental Protection Agency (U.S. EPA). Ambient Water Quality Criteria Recommendations (Information Supporting the Development of State and Tribal Nutrient Criteria): Rivers and Streams in Nutrient Ecoregion VI, EPA 822-B-00-017; U.S. EPA Office of Water: Washington, DC, 2000.

Walbridge, M.A.; Struthers, J.P. Phosphorus Retention in Non-Tidal Palustrine Forested Wetlands of the Mid-Atlantic Region. *Wetlands* **1993**, *13*, 84–94.

Water Environment Research Foundation. Using WERF to Cut Water Costs. *Progress*. 2004. http://www.werf.org/press/summer01/01su_energycosts.cfm

Weyer, P.J.; Cerhan, J.R.; Kross, B.C.; Hallberg, G.R.; Kantamneni, J.; Breuer, G.; Jones, M.P.; Zheng, W.; Lynch, C.F. Municipal Drinking Water Nitrate Level and Cancer Risk in Older Women: The Iowa Health Study. *Epidemiology* **2001**, *12*, 327–338.

Zenz, D.R. *Technical Feasibility and Cost to Meet Nutrient Standards in the State of Illinois*. Report prepared for Illinois Association of Wastewater Agencies, Springfield, Ill., 2003.

WASTEWATER UTILITY

Alabama
Montgomery Water Works & Sanitary Sewer Board

Alaska
Anchorage Water & Wastewater Utility

Arizona
Gila Resources
Glendale, City of, Utilities Department
Mesa, City of
Peoria, City of
Pima County Wastewater Management

Arkansas
Little Rock Wastewater Utility

California
Calaveras County Water District
Central Contra Costa Sanitary District
Crestline Sanitation District
Delta Diablo Sanitation District
Dublin San Ramon Services District
East Bay Dischargers Authority
East Bay Municipal Utility District
Eastern Municipal Water District
El Dorado Irrigation District
Fairfield-Suisun Sewer District
Fort Lauderdale, City of
Irvine Ranch Water District
Las Virgenes Municipal Water District
Livermore, City of
Lodi, City of
Los Angeles, City of
Los Angeles County, Sanitation Districts of
Napa Sanitation District
Orange County Sanitation District
Palo Alto, City of
Riverside, City of
Sacramento Regional County Sanitation District
San Diego Metropolitan Wastewater Department, City of
San Francisco, City & County of
San Jose, City of
Santa Barbara, City of
Santa Cruz, City of
Santa Rosa, City of
South Bayside System Authority
South Coast Water District
South Orange County Wastewater Authority
Sunnyvale, City of
Union Sanitary District
West Valley Sanitation District

Colorado
Boulder, City of
Colorado Springs Utilities
Greeley, City of
Littleton/Englewood Water Pollution Control Plant
Metro Wastewater Reclamation District, Denver

Connecticut
The Mattabassett District
New Haven, City of, WPCA

District of Columbia
District of Columbia Water & Sewer Authority

Florida
Broward, County of
Fort Lauderdale, City of
Gainesville Regional Utilities
JEA
Miami-Dade Water & Sewer Authority
Orange County Utilities Department
Reedy Creek Improvement District
Seminole County Environmental Services
St. Petersburg, City of
Stuart Public Utilities
Tallahassee, City of
Tampa, City of
Toho Water Authority
West Palm Beach, City of

Georgia
Atlanta Department of Watershed Management
Augusta, City of
Clayton County Water Authority
Cobb County Water System
Columbus Water Works
Fulton County
Gwinnett County Department of Public Utilities
Savannah, City of

Hawaii
Honolulu, City & County of

Idaho
Boise, City of

Illinois
American Bottoms Wastewater Treatment Plant
Greater Peoria Sanitary District
Kankakee River Metropolitan Agency
Metropolitan Water Reclamation District of Greater Chicago
Wheaton Sanitary District

Iowa
Ames, City of
Des Moines Metro Wastewater Reclamation Authority
Iowa City

Kansas
Johnson County Unified Wastewater Districts
Unified Government of Wyandotte County/ Kansas City, City of

Kentucky
Louisville & Jefferson County Metropolitan Sewer District

Louisiana
Sewerage & Water Board of New Orleans

Maine
Bangor, City of
Portland Water District

Maryland
Anne Arundel County Bureau of Utility Operations
Howard County Department of Public Works
Washington Suburban Sanitary Commission

Massachusetts
Boston Water & Sewer Commission
Upper Blackstone Water Pollution Abatement District

Michigan
Ann Arbor, City of
Detroit, City of
Holland Board of Public Works
Lansing, City of
Saginaw, City of
Wayne County Department of Environment
Wyoming, City of

Minnesota
Rochester, City of
Western Lake Superior Sanitary District

Missouri
Independence, City of
Kansas City Missouri Water Services Department
Little Blue Valley Sewer District
Metropolitan St. Louis Sewer District

Nebraska
Lincoln Wastewater System

Nevada
Henderson, City of

New Jersey
Bergen County Utilities Authority
Passaic Valley Sewerage Commissioners

New York
New York City Department of Environmental Protection

North Carolina
Charlotte/Mecklenburg Utilities
Durham, City of
Metropolitan Sewerage District of Buncombe County
Orange Water & Sewer Authority

Ohio
Akron, City of
Butler County Department of Environmental Services
Columbus, City of
Metropolitan Sewer District of Greater Cincinnati
Northeast Ohio Regional Sewer District
Summit, County of

Oklahoma
Trinity River Authority
Tulsa, City of

Oregon
Clean Water Services
Eugene/Springfield Water Pollution Control
Water Environment Services

Pennsylvania
Philadelphia, City of
University Area Joint Authority

South Carolina
Charleston Commissioners of Public Works
Mount Pleasant Waterworks & Sewer Commission
Spartanburg Sanitary Sewer District

Tennessee
Cleveland, City of
Knoxville Utilities Board
Murfreesboro Water & Sewer Department
Nashville Metro Water Services

Texas
Austin, City of
Dallas Water Utilities
Denton, City of
El Paso Water Utilities
Fort Worth, City of
Gulf Coast Waste Disposal Authority
Houston, City of
San Antonio Water System
Trinity River Authority

Utah
Salt Lake City Corporation

Virginia
Alexandria Sanitation Authority
Arlington, County of
Fairfax County
Hampton Roads Sanitation District
Henrico, County of

Hopewell Regional Wastewater Treatment Facility
Loudoun County Sanitation Authority
Lynchburg Regional WWTP
Prince William County Service Authority
Richmond, City of
Rivanna Water & Sewer Authority

Washington
Bellevue Utilities Department
Edmonds, City of
Everett, City of
King County Department of Natural Resources
Seattle Public Utilities
Sunnyside, Port of
Yakima, City of

Wisconsin
Green Bay Metro Sewerage District
Madison Metropolitan Sewerage District
Milwaukee Metropolitan Sewerage District
Racine, City of
Sheboygan Regional Wastewater Treatment
Wausau Water Works

Australia
South Australian Water Corporation
Sydney Water Corporation
Water Corporation of Western Australia

Canada
Greater Vancouver Regional District
Regina, City of, Saskatchewan
Toronto, City of, Ontario
Winnipeg, City of, Manitoba

Mexico
Servicios de Agua y Drenaje de Monterrey, I.P.D.

New Zealand
Watercare Services Limited

United Kingdom
Yorkshire Water Services Limited

STORMWATER UTILITY

California
Los Angeles, City of, Department of Public Works
Monterey, City of
San Francisco, City & County of
Santa Rosa, City of

Colorado
Boulder, City of

Georgia
Griffin, City of

Iowa
Cedar Rapids Wastewater Facility
Des Moines Metro Wastewater Reclamation Authority

Kansas
Overland Park, City of

Maine
Portland Water District

North Carolina
Charlotte, City of, Stormwater Services

Pennsylvania
Philadelphia, City of

Tennessee
Chattanooga Stormwater Management

Washington
Seattle Public Utilities

STATE

Arkansas Department of Environmental Quality
Fresno Department of Public Utilities
Kansas Department of Health & Environment
Urban Drainage & Flood Control District, CO

CORPORATE

ADS Environmental Services
The ADVENT Group Inc.
Alan Plummer & Associates
Alden Research Laboratory Inc.
Aqua-Aerobic Systems Inc.
Aquateam–Norwegian Water Technology Centre A/S
ARCADIS
Associated Engineering
BaySaver Inc.
Black & Veatch
Blasland, Bouck & Lee Inc.
Boyle Engineering Corporation
Brown & Caldwell
Burns & McDonnell
CABE Associates Inc.
The Cadmus Group
Camp Dresser & McKee Inc.
Carollo Engineers Inc.
Carpenter Environmental Associates Inc.
CDS Technologies Inc.
Chemtrac Systems Inc.
CH2M HILL
Damon S. Williams Associates, LLC
David L. Sheridan, P.C.
Dewling Associates Inc.
Earth Tech Inc.
Ecovation
EMA Inc.
The Eshelman Company Inc.
EWT Holdings Corporation
Fay, Spofford, & Thorndike Inc.
Freese & Nichols Inc.
ftn Associates Inc.
Fuss & O'Neill Inc.
Gannett Fleming Inc.
GE Betz
Greeley and Hansen LLC
Hazen & Sawyer, P.C.
HDR Engineering Inc.
HNTB Corporation
HydroQual Inc.
Infilco Degremont Inc.
Ingersoll-Rand Energy Systems
Insituform Technologies Inc.
Institute for Environmental Technology & Industry, Korea
Jacobson Helgoth Consultants Inc.
Jason Consultants LLC Inc.
Jordan, Jones, & Goulding Inc.
KCI Technologies Inc.
Kelly & Weaver, P.C.
Kennedy/Jenks Consultants
Komline Sanderson Engineering Corporation
Lawler, Matusky & Skelly Engineers, LLP
LTI Inc.
Lombardo Associates Inc.
Malcolm Pirnie Inc.
McKim & Creed
MEC Analytical Systems Inc.
Metcalf & Eddy Inc.
MWH
New England Organics
O'Brien & Gere Engineers Inc.
Odor & Corrosion Technology Consultants Inc.
Oswald Green, LLC
PA Government Services Inc.
Parametrix Inc.
Parsons
Post, Buckley, Schuh & Jernigan
R&D Engineering/Conestoga Rover & Associates
RMC Inc.
R.M. Towill Corporation
Ross & Associates Ltd.
Royce Technologies
SAIC Maritime Technical Group
Short Elliott Hendrickson Inc.
Stantec Consulting Inc.
Stearns & Wheler, LLC
Stone Environmental Inc.
Stormwater Management Inc.
Stratus Consulting Inc.
Synagro Technologies Inc.
Tetra Tech Inc.
Trojan Technologies Inc.
URS Corporation
USfilter
Veola Water North America
Wade-Trim Inc.
Westin Engineering Inc.
Weston Solutions Inc.
Woodard & Curran
WRc/D&B, LLC
WWETCO, LLC
Zenon Environmental Inc.
Zoeller Pump Company

INDUSTRY

American Electric Power
ChevronTexaco Energy Research & Technology Company
The Coca-Cola Company
Dow Chemical Company
DuPont Company
Eastman Chemical company
Eastman Kodak Company
Eli Lilly & Company
Merck & Company Inc.
ONDEO Services
Procter & Gamble Company
PSEG Services Corporation
RWE Thames Water Plc
Severn Trent Services Inc.
United Water Services LLC

Note: List as of 2/1/05

WERF Product Order Form

As a benefit of joining the Water Environment Research Foundation, subscribers are entitled to receive one complimentary copy of all final reports and other products. Additional copies are available at cost (usually $10). To order your complimentary copy of a report, please write "free" in the unit price column. WERF keeps track of all orders. If the charge differs from what is shown here, we will call to confirm the total before processing.

Name _______________ Title _______________

Organization _______________

Address _______________

City _______________ State _______ Zip Code _______ Country _______

Phone _______________ Fax _______________ Email _______________

Stock #	Product	Quantity	Unit Price	Total
			Postage & Handling	
			VA Residents Add 4.5% Sales Tax	
			Canadian Residents Add 7% GST	
			TOTAL	

Method of Payment: (All orders must be prepaid.)

❑ Check or Money Order Enclosed

❑ Visa ❑ Mastercard ❑ American Express

Account No. _______________ Exp. Date _______

Signature _______________

Shipping & Handling:

Amount of Order	United States	Canada & Mexico	All Others
Up to but not more than:	Add:	Add:	Add:
$20.00	$5.00*	$8.00	50% of amount
30.00	5.50	8.00	40% of amount
40.00	6.00	8.00	
50.00	6.50	14.00	
60.00	7.00	14.00	
80.00	8.00	14.00	
100.00	10.00	21.00	
150.00	12.50	28.00	
200.00	15.00	35.00	
More than $200.00	Add 20% of order	Add 20% of order	

*minimum amount for all orders

Note: Please make checks payable to the Water Environment Research Foundation.

To Order (Subscribers Only):

Log on to www.werf.org and click on the "Product Catalog."

Phone: (703) 684-2470
Fax: (703) 299-0742.

WERF
Attn: Subscriber Services
635 Slaters Lane
Alexandria, VA 22314-1177

To Order (Non-Subscribers):

Non-subscribers may be able to order WERF publications either through WEF (www.wef.org) or IWAP (www.iwapublishing.com). Visit WERF's website at www.werf.org for details.

IWA Publishing's authorised EU representative for General Product Safety Regulations is Diane D'Arras, 15 rue Duret, 75116 Paris, France, e-mail: safety@iwap.co.uk.

Printed and bound by CPI Group (UK) Ltd, Croydon, CR0 4YY
17/09/2026
02218798-0005